NUMBERPLAY

NUMBER-±PLAY÷×

GYLES BRANDRETH

Rawson Associates : New York

ISBN O 89256 267 6

First published in the United States of America in 1984 by
Rawson Associates, 597 Fifth Avenue, New York, N.Y. 10017
First published in Great Britain in 1984 by Severn
House Publishers Limited, 4 Brook Street, London
W1Y 1AA.

Typeset by Gee Graphics, 15/27 Gee Street, London,
EC1.

Printed and bound in Great Britain by Anchor
Brendon Ltd, Tiptree, Essex.

CONTENTS

FOREWORD

There was a young Fellow of Trinity
Who found the square root of infinity
But the number of digits
Gave him the fidgets;
So he dropped Maths and took up Divinity.

This is a book by a man who almost failed O Level maths and abandoned university economics at the end of his first year. Fortunately that doesn't matter, because this isn't a book about mathematics or economics – or divinity for that matter. It's a book about *numbers* – in all their multifarious manifestations, from telephone numbers to 'magic squares' of numbers, from the number of days in the week to the number of dollars in a sextillionaire's bank account.

'One has to be able to count,' wrote Maxim Gorky, 'if only so that at 50 one doesn't marry a girl of 20.' Quite so. If you happen to own a calculator (or eight fingers and two thumbs) and can count – and perhaps do a little adding, subtracting, multiplying and dividing on the side – all the puzzles and challenges in the book will be within your grasp.

I figured my readers would already know that the cube of 99 is 970299 and that the denary equivalent of binary 0·000001 is 0·015625 (2^{-6}), so what I have tried to do in this A to Z of numerical facts and fancies is to provide you with information not already at your fingertips.

To me the joy of numbers is that they are so full of surprises.

AMAZING NUMBERS

DID YOU KNOW THAT ...

... If you wanted to arrange 15 books on a shelf in every possible way, and you made one change every minute, it would take you 2,487,996 years to do it?

... One pound of iron contains an estimated 4,891,500,000,000,000,000,000,000 atoms?

... When three pairs of rabbits were released in Australia in the middle of the nineteenth century, it took less than a decade for them to multiply into millions, threatening the Australian agricultural economy?

... Every cubic mile of sea water contains 170,000,000 tons of chemical compounds, not counting the water itself?

... There are some 318,979,564,000 possible ways of playing the first four moves on each side in a game of chess?

... The earth travels over one and a half million miles every day?

... Scientists reckon that the highest temperature in the universe is probably 120,000 times greater than the interior temperature of the Sun (14,000,000°C)?

... Almost one person in four on Earth is Chinese and some estimates put the population of China by the end of the century at 1,800,000,000?

. . . There are 2,500,000 rivets in the Eiffel Tower?

. . . 500 sets of the 1112-volume edition of British Parliamentary Papers used £15,000 worth of gold and the skins from 34,000 Indian goats? Each set weighed over three tons.

. . . If all the blood vessels in the human body were laid end to end, they would stretch for 100,000 miles?

. . . If you were to write down all the different ways of arranging our 26 letter alphabet on a strip of paper, you would need a strip 160 million light-years long?

. . . When you want to express either very large or very small numbers it is usual to write them in this compact form:

$$10^{10} = 10,000,000,000$$
$$10^{9} = 1,000,000,000$$
$$10^{8} = 100,000,000$$
$$10^{7} = 10,000,000$$
$$10^{6} = 1,000,000$$
$$10^{5} = 100,000$$
$$10^{4} = 10,000$$
$$10^{3} = 1,000$$
$$10^{2} = 100$$
$$10^{1} = 10$$
$$10^{0} = 1$$
$$10^{-1} = 0.1$$
$$10^{-2} = 0.01$$
$$10^{-3} = 0.001$$
$$10^{-4} = 0.0001$$
$$10^{-5} = 0.00001$$
$$10^{-6} = 0.000001$$
$$10^{-7} = 0.0000001$$
$$10^{-8} = 0.00000001$$
$$10^{-9} = 0.000000001$$
$$10^{-10} = 0.0000000001$$

. . . Words dealing with very high numbers have been in use for far longer than many people imagine? The word *million* first came into use in English in 1370 and *trillion* has been around since the 1480s. The first British use of words dealing with even larger numbers is as follows:

quadrillion 1674 $= 10^{24}$
quintillion 1674 $= 10^{30}$
sextillion 1690 $= 10^{36}$
septillion 1690 $= 10^{42}$
octillion 1690 $= 10^{48}$
nonillion 1828 $= 10^{54}$
decillion 1845 $= 10^{60}$

. . . There was a family in France called 1792? The family had four sons each named after a month: January 1792, February 1792, March 1792 and April 1792.

. . . There are approximately 54,000,000,000,000,000,000,000,000,000 different bridge deals?

. . . The earth only receives 1/2,000,000,000th of the sun's energy?

. . . If the world's top 20 mountains were laid end to end westwards from London, the peak of the last one would be only a few miles short of Gloucester, 103 miles away?

. . . Assuming that you can count to 200 in one minute, and that you decide to count for 12 hours a day at this rate, and that you don't have any pressing engagements in the foreseeable future, it would take you in the region of 19,024 years, 68 days, 10 hours and 40 minutes to count to one billion?

. . . Protons are stable for 10^{28} years, but neutrons only last for 15 minutes. If you compared the lifespan of a neutron with the average time it takes to drink a pint of beer, you would need to drink for 20,000,000,000,000 times the lifespan of the Universe to last as long as a proton?

. . . If you dismantled the Great Pyramid of Cheops and used the stone to build a wall 20 inches high and one brick thick, the wall would stretch all the way round the earth?

. . . If you scaled the nucleus of an atom up to the size of a marble, half-an-inch in diameter, it would have a mass of 100,000,000 tons?

. . . You could arrange the 52 cards in an ordinary deck of playing cards about 80,660,000,000,000,000, 000,000,000,000,000,000,000,000,000,000,000,000, 000,000,000,000,000 different ways?

. . . The largest number you can write with three digits is $9(9^9)$ which is nine raised to its 387,420,489th power? It's estimated that the row of digits would be about 600 miles long and would take about 150 years to read.

. . . One estimate states that collectively the population of the USA is carrying about 2,000,000 tons of excess fat?

. . . Almost one ballpoint pen per head of population was sold in the United Kingdom during the first year of their appearance on the UK market? The total was some 53,000,000.

. . . If you decided to place one penny on the first square of a chess board, followed by two pennies on the second square, four on the third, eight on the fourth and so on, doubling the number of pennies as you progressed, by the time you reached the last square (64th) you would have a pile of 9,223,372,036,854,775,808 pennies?

BY HOW MUCH?

MEASURED TO SCALE

You can find scales almost wherever you turn – on a fish, in the bathroom, on the piano – but the scales I find the most intriguing are the less familiar ones that follow. The first two govern a couple of the senses: sound and touch.

Sound
Decibels measure the relative loudness or intensity of sounds. A 20 decibel sound is ten times lounder than a sound that measures 10 decibels, a 30 decibel sound is a hundred times louder, and so on. Approximately one decibel has been established as the smallest difference between sounds that can be registered by the human ear. For easy reckoning, this scale should help you gauge how loud a noise is:

10 decibels	A light whisper
20 decibels	Quiet conversation
30 decibels	Normal conversation
40 decibels	Light traffic
50 decibels	Typewriter, loud conversation
60 decibels	Noisy office
70 decibels	Normal traffic, quiet train
80 decibels	Rock music, underground train

90 decibels	Heavy traffic, thunder
100 decibels	Jet aircraft at take-off
140 decibels	OW! (Painful)

Touch
Moh's hardness scale
This works on the principle of comparing the hardness of materials by measuring them with ten standard minerals, as follows:

Mineral	Hardness Test	Moh's Hardness
Talc	Crushed by fingernail	1.0
Gypsum	Scratched by fingernail	2.0
Calcite	Scratched by copper coin	3.0
Fluorspar	Scratched by glass	4.0
Apatite	Scratched by penknife	5.0
Feldspar	Scratched by quartz	6.0
Quartz	Scratched by steel file	7.0
Topaz	Scratched by corundum	8.0
Corundum	Scratched by diamond	9.0
Diamond	Scratched by diamond	10.0

And here are a couple of scales for measuring the elemental forces.

Earthquakes
Today we're used to hearing the force of earthquakes quoted on the Richter Scale, but there are other scales for measuring earthquakes. My favourite is the Mercalli Intensity Scale. It measures the force of an earthquake by *impression* rather than by the use of instruments – which is probably more useful if you happen to be standing on top of an earthquake when it starts and don't have your pocket seismograph handy.

Mercalli Intensity Scale
I   Just detectable by experienced observers when prone. Microseisms.

II   Felt by few. Delicately poised objects may sway.
III Vibration but still unrecognized by many. Feeble.
IV   Felt by many indoors but by few outdoors. Moderate.
V   Felt by most all. Many awakened. Unstable objects move.
VI   Felt by all. Heavy objects move. Alarm. Strong.
VII General alarm. Weak buildings considerably damaged. Very strong.
VIII Damage general except in proofed buildings.
IX   Buildings shifted from foundations collapse, ground cracks. Highly destructive.
X   Masonry buildings destroyed, rails bent, serious ground fissures. Devastating.
XI   Few if any structures left standing. Bridges down. Rails twisted. Catastrophic.
XII Damage total. Vibrations distort vision. Objects thrown in air. Major catastrophe.

Winds

The Beaufort Scale is a series of numbers from 0 to 17 that designate the force of the wind. The numbers 0 to 12 were arranged by Admiral Sir Francis Beaufort in 1806 to indicate the strength of the wind from a calm, force 0, to a hurricane, force 12 – 'that which no canvas can withstand.' The Beaufort Scale numbers 13 to 17 were added by the United States Weather Bureau in 1955.

Beaufort Number	Miles per hour	Description	Observation
0	0-1	Calm	Smoke rises vertically
1	1-3	Light Air	Smoke drifts slowly
2	4-7	Slight breeze	Leaves rustle

Beaufort Number	Miles per hour	Description	Observation
3	8-12	Gentle breeze	Leaves and twigs rustle
4	13-18	Moderate breeze	Small branches move
5	19-24	Fresh breeze	Small trees sway
6	25-31	Strong breeze	Large branches sway
7	32-38	Moderate gale	Whole trees move
8	39-46	Fresh gale	Twigs break off trees
9	47-54	Strong gale	Branches break
10	55-63	Whole gale	Trees snap and are blown down
11	64-75	Storm	Widespread damage
12	More than 75	Hurricane	Extreme damage
13	83-92	} Hurricane	
14	93-103		
15	104-114		
16	115-125		
17	126-136		

Acidimeters to Zymometers

Acidimeters and Zymometers are the alpha and the omega of measuring instruments. As you rightly guessed, an acidimeter measures acidity. And in case you didn't know already, a zymometer is an instrument used to measure fermentation. Here's a selection of 20 unlikely-meters with details of what they measure.

16

1	Craniometer	Skulls
2	Halometer	Form, angles and planes of crystals
3	Sillometer	Speed of ships
4	Viameter	Distance travelled
5	Oscillometer	Roll of ships or spacecraft
6	Galactometer	Flow of milk
7	Oometer	Birds' eggs
8	Telemeter	Distant objects
9	Astrophometer	Intensity of star's light
10	Clinometer	Angles of elevation
11	Dendrometer	Trees
12	Pyrheliometer	Sun's heat
13	Litrameter	Specific gravities in liquid
14	Atmometer	Rates of exhalation of moisture
15	Inclinometer	Terrestial magnetic force
16	Konometer	Dust
17	Tribometer	Sliding friction
18	Drosometer	Dew
19	Orometer	Height of mountains
20	Reflectometer	Smoke content

Now it's your turn. Below are ten other instruments with the materials, elements or phenomena they measure. Unfortunately they're in the wrong order. See if you can pair them correctly. (Answers at the back.)

1	Gradiometer	a) Small electric currents
2	Comptometer	b) Calculation
3	Galvanometer	c) Salinity
4	Potometer	d) Water intake
5	Salimeter	e) Specific gravity of urine
6	Urinometer	f) Gradients
7	Argentometer	g) Lung capacity
8	Polarimeter	h) Strength of silver solutions
9	Vibrometer	i) Vibrations
10	Spirometer	j) Polarization of light

Seconds away

An important year in the life of the humble second was 1967. It was in that year that the second was re-defined as 9,192,631,770 periods of caesium vibration.

As far as most people are concerned, it's enough to know there are 60 seconds in a minute. However, if your brain can cope with it, you may want to explore the realms of the gigasecond or investigate the tiny lifespan of the attosecond. (For the uninitiated a gigasecond is just over 30 years and an attosecond, abbreviated to *as*, is equal to 0.000000000000000001 of a second). You certainly ought to know that there are 86,400 seconds in a day, 604,800 seconds in a week, 2,592,000 seconds in a 30-day month and that one second is 0.0000116 of a day.

Before getting too carried away with the ramifications of measuring time with this degree of accuracy, it's worth pondering on the opinion of the scientist who was asked which was better, a broken watch or one that ran ten seconds slow per day. After due consideration, the scientist announced that the broken watch was certainly the better watch, since it would show the right time twice a day, while the watch that was ten seconds slow would only be accurate once in 11.8 years.

MEASURES MISCELLANY

Many early measures have long since fallen from use while others have application for only a small number of specialist activities. Here are 20 units and measures and the 20 uses to which they are put. How many of them can you match up correctly? (Answers at the back.)

1	Balthazer	a)	Inductance
2	Farad	b)	Force

3	Quire	c)	Electrical resistance
4	Ampere	d)	24 sheets of paper
5	Link	e)	Eight magnums of champagne
6	Stadium	f)	Thermodynamic pressure
7	Radian	g)	120 fathoms
8	Hertz	h)	Hundredth part of a chain
9	Henry	i)	Electric current
10	Jeroboam	j)	Frequency
11	Newton	k)	Power
12	Coulomb	l)	Electrical capacitance
13	Joule	m)	Work, energy
14	Watt	n)	6.4 pints of champagne/brandy
15	Kelvin	o)	202 yards
16	Ohm	p)	Plane angle
17	Ream	q)	480 sheets of paper
18	Hand	r)	Electric charge
19	Nautical mile (International)	s)	Four inches
20	Cable	t)	6076.1033 feet

CALCULATOR CAPERS

THE COMMUNICATING CALCULATOR

Calculators can be charmingly communicative instruments, given a chance and a bit of a helping hand. Are you fed up calculating your VAT, or trying to work out how much worse off you are now than you were a year ago? You are? Well, your calculator can console you. Enter 0.04008 into the calculator, turn it upside down and read what is shown on the screen. There you are! Your calculator is showing proper fellow feeling; it's saying, 'Boohoo'.

Limited as their vocabulary is, calculators can spell out a surprising number of words, all using nine letters of the alphabet that are obtained thus:

To get B enter 8
To get E enter 3
To get G enter 6 (for a capital G enter 9)
To get H enter 4
To get I enter 1
To get L enter 7
To get O enter 0
To get S enter 5
To get Z enter 2

Remember to enter your numbers into the calculator in reverse order, since you'll be reading the 'letters' upside down. Clearly a dictionary – or a marvellous book I know called *Wordplay** – will provide you with the vocabulary you need, but to get you started, here are a few basic words with the numbers you have to enter in order to obtain them:

To spell BEE enter 338
To spell BEG enter 638
To spell BLESS enter 55378
To spell BOBBLE enter 378808
To spell BOGGLE enter 376608
To spell BOIL enter 7108
To spell EGG enter 993
To spell GEESE enter 35336
To spell GIGGLE enter 379919
To spell GLEE enter 3376
To spell GOBBLE enter 378806
To spell GOSH enter 4506
To spell HE enter 34
To spell HILL enter 7714
To spell HISS enter 5514
To spell HOSE enter 3504
To spell ILL enter 771
To spell LESS enter 5537
To spell LOG enter 607
To spell LOOSE enter 35007
To spell LOSE enter 3507
To spell LOSS enter 5507
To spell OBOE enter 3080
To spell SEIZE enter 32135
To spell SELL enter 7735
To spell SHELL enter 77345
To spell SHOE enter 3045

*Published in the United States as *The Joy of Lex*.

To spell SIEGE enter 36315
To spell SIGH enter 4615
To spell SIZE enter 3215
To spell SLOSH enter 45075
To spell SOB enter 805

CALCULATOR POKER

When you've run out of calculated conversation, you can play cards with your calculator instead.

In this Calculator version of Poker, the aim of the game is exactly the same as it is in the original card game, namely to end up with the best hand. The chart below gives you the hands. Apart from that, all you need are two or more players, one calculator per player, paper and pencils.

The game proceeds like this:

1 Each player enters a random four-digit number into his or her calculator. Zero is not allowed.
2 Each player presses the X key and then passes the calculator to the left. The screens must be covered when they are passed.
3 Without looking at the numbers on the screens, the players now enter another random, four-digit number, zero excepted, and pass the calculators back to their owners.
4 The owners press the equals key and examine the answers.
5 The playing hand is formed by the seven digits on the right of the screen. From this choice, each player must select the five digits that will give him or her the best hand. (In this game zero counts as ten, and therefore is the highest score.)
6 When all the players have selected their hands they look at the chart and decide who has the best hand, and is therefore the winner. The highest scoring

hands are listed first. Winners are awarded one point. The first player to get ten points wins the game.

The hands

Five of the same kind	eg. 88888
Straight flush – even	eg. 24680
Straight flush – odd	eg. 51379
Four of the same kind	eg. 63666
Full House	eg. 47447
Flush – even	eg. 24648
Flush – odd	eg. 13351
Straight	eg. 12345
Three of the same kind	eg. 98995
Two pairs	eg. 48184
One pair	eg. 73274

If there is a tie, the player with the highest number wins.

OUT IN SIX

The object of this calculator game is to reach zero in just six goes, and you won't be competing with anyone else, because this is a game for you to play by yourself. This is what you do:

1 Enter any six-digit number into your calculator, providing that you do not use zero and that no digit is repeated.
2 You now have six goes to try and reach zero.
3 You can use any mathematical function you like in each go and you can use any two-digit number.
4 Although it might seem obvious to keep dividing and subtracting to reduce your total, you may find it better to add or even multiply to reach a number that can then be cleanly divided by another.

(Once you've mastered bringing it down to zero in six
goes, why not try doing it in five?)

Example

1 Suppose you enter 583621		583621
2 In your first move you add 19	+	19
		583640
Then you divide by 40	=	14591
Then you subtract 11	=	14580
Then you divide by 45	=	324
Then you divide by 12	=	27
Then you subtract 27 from the last go	=	0

ODDS AND EVENS

This game uses eight digits – 1,2,3,4,6,7,8,9 – split into
odds and evens. One player takes the odd numbers,
the other takes the evens and they each try to reach a
higher total than the other. This is how the game
progresses:

1 Toss to see who will be odd. Odd goes first.
2 Both players choose a number and enter it into their
respective calculators.
3 Odd now calls out one mathematical function.
Multiplication may only be called *once* by each
player, so you must be careful when to call it.
4 Both players press the function key called and then
enter another of their digits. After this they each
press the = key.
5 Even now calls outs one of the mathematical
functions. The players enter this and then selects
another of their digits. They then press the = key.
6 Play continues like this until the digits are used up.

The player with the highest total is the winner.
Here's a sample game:

Odd enters		7	Even enters		6
Calls ADD					
Enters 3	=	10	Enters 4	=	10
			Calls MULTIPLY		
Enters 9	=	90	Enters 8	=	80
Calls DIVIDE					
Enters 1	=	90	Enters 2	=	40

Odd wins with the highest total.

TOP TEN

For this game you'll need a pair of dice as well as one calculator per person. Each player takes a turn at rolling the dice, and according to the total shown on them, will perform one of the four mathematical functions on his calculator. The object of the game is to finish with the highest score after ten rounds.
You play like this:

1 Toss a coin to decide who will start.
2 The players take turns in rolling the dice and enter the totals they roll into their own calculators.
3 The first player now rolls the dice again and looks at the table to see what he or she should do. If he or she:

 . . . rolls 2, add 2
 . . . rolls 3, enter 3 (no function)
 . . . rolls 4, multiply by 4
 . . . rolls 5, subtract 5
 . . . rolls 6, add 6
 . . . rolls 7, multiply by 7
 . . . rolls 8, subtract 8
 . . . rolls 9, add 9
 . . . rolls 10, enter 10 (no function)
 . . . rolls 11, divide by 11
 . . . rolls 12, multiply by 12

4 The second player then rolls the dice and looks at the chart to see what to do.

5 Play continues likes this until all the players have rolled the dice ten times. The player with the highest total is the winner.

Sample game

Player A		Player B		Player C	
rolls 6	6	rolls 9	9	rolls 11	11
rolls 5 (−)	1	rolls 4 (×)	36	rolls 7 (×)	77
rolls 4 (×)	4	rolls 2 (+)	38	rolls 9 (+)	86
rolls 9 (+)	13	rolls 3 (/)	383	rolls 5 (−)	81
rolls 8 (−)	5	rolls 2 (+)	385	rolls 10 (/)	8110
rolls 12 (×)	60	rolls 4 (×)	1540	rolls 3 (/)	81103
rolls 6 (+)	66	rolls 7 (×)	10780	rolls 9 (+)	81112
rolls 2 (+)	68	rolls 5 (−)	10775	rolls 11 (÷)	7373.8
rolls 12 (×)	816	rolls 11 (÷)	979.5	rolls 8 (−)	7365.8
rolls 4 (×)	3264	rolls 5 (−)	974.5	rolls 5 (−)	7360.8

Player C wins with the highest total.

QUIZCULATION

These eight instructions combine a quiz and a calculation to test both you and your calculator. Simply answer the questions, follow the instructions and check your answer at the back of the book.

1 Multiply the date of the Battle of Stamford Bridge by a 'bakers's dozen'.

2 Multiply the answer by the number of days in the year when the last Olympic Games were held.

3 Now subtract the number of days in an ordinary February.

4 Divide the answer by the number of years in a millenium.

5 Subtract the number of wives of Henry VIII.

6 Subtract the first year of William I's reign in England.

7 Divide the total by the number of years a centenarian has lived.

8 Subtract the steps in a famous spy novel by John Buchan – what's your answer?

HOW OLD ARE YOU?

How old are you? That's not a question that one can put to anyone, which is why I am only asking the youngsters. If you were born on or after 1 January 1960, how old are you? No, not in years! Not in months, not in days, not in hours, not in minutes, but in *seconds*! How many seconds have gone by since you were born? With a calculator you can work out the answer. Here's how.

Step 1 Copy the list of years below. Draw a line under the year you were born and a second line above the present year. Count the number of years between the two lines you have drawn, and multiply this number by 365. Then count the number of Xs between the two lines, and add that number to the calculator total.

1960	×	1973	
1961		1974	
1962		1975	
1963		1976	×
1964	×	1977	
1965		1978	
1966		1979	
1967		1980	×
1968	×	1981	
1969		1982	
1970		1983	
1971		1984	×
1972	×	1985	

Step 2 In the list of months below find the month you

were born in, and add the number next to it to the previous calculator total.

January	366	July	184
February	335	August	153
March	306	September	122
April	275	October	92
May	245	November	61
June	214	December	31

Step 3 Subtract the day of the month on which you were born.

Step 4 If you were born in January or February of any year not marked X, subtract one. If you were born in March or any later month, no matter what year it was, do nothing in this step. And if you were born at any time in any of the years marked X, also do nothing.

Step 5 Using the new list of months below, find the present calendar month and add the number next to it to the previous total.

January	0	July	182
February	31	August	213
March	60	September	244
April	91	October	274
May	121	November	305
June	152	December	335

Step 6 If the current month is not January or February and the current year is not marked X, subtract one. If the month is not January or February, but the year is marked X, do not subtract one. And if the year is not marked X, but the month is January or February, again do nothing in this step.

Step 7 Add one number less than the day of the month it is today.

Step 8 Multiply by 24.

Step 9 If you do not know the time of the day you were born, add 24 to the calculator total, skip steps **9** and **10**, and go on to **Step 11**. Your answer will not be exact, but it will still be very close. If you do know the time, take the hour you were born and find it in the list of hours below. Then add the number next to it to your calculator total.

Midnight	24	Noon	12
1 am	23	1 pm	11
2 am	22	2 pm	10
3 am	21	3 pm	9
4 am	20	4 pm	8
5 am	19	5 pm	7
6 am	18	6 pm	6
7 am	17	7 pm	5
8 am	16	8 pm	4
9 am	15	9 pm	3
10 am	14	10 pm	2
11 am	13	11 pm	1

Step 10 See what time it is now. Again ignore the minutes. Find the hour in the list of hours below and add the number given next to it to your present calculator total.

Midnight	0	Noon	12
1 am	1	1 pm	13
2 am	2	2 pm	14
3 am	3	3 pm	15
4 am	4	4 pm	16
5 am	5	5 pm	17
6 am	6	6 pm	18
7 am	7	7 pm	19
8 am	8	8 pm	20
9 am	9	9 pm	21
10 am	10	10 pm	22
11 am	11	11 pm	23

Step 11 Multiply by 60
Step 12 If you do not know the minute you were born,

skip steps 12 and 13 and go on to step 14. If you do know the minute, ignore the hour, and subtract the number of minutes from the calculator total.

Step 13 See what time it is. Ignoring the hour, add the minutes to the calculator total.

Step 14 This is your up-to-the-minute age in minutes. To work out your age in seconds, you will have to multiply by 60. If your calculator only has an eight-digit display you can get the answer by multiplying by six and then writing the calculator total on paper with an extra zero at the end.

So now you know the answer, how does it feel to be millions of seconds old?

DIGITAL DELIGHTS

WHAT'S A DIGIT?

To an anatomist a digit is a finger or a toe. To an astronomer measuring an eclipse, a digit is a twelfth part of a sun's or a moon's diameter.

To the numerate a digit is any number from zero to nine.

Nine-digit addition

In each of these sums the nine digits from one to nine are used once, and once only.

243	341	154	317
+675	+586	+782	+628
918	927	936	945
216	215	318	235
+738	+748	+654	+746
954	963	972	981

The totals are intriguing: each begins with a nine and is followed by a figure that is a multiple of nine.

Squaring up

Have you any idea what is the smallest number with a square that contains all the digits from zero to nine used once? And have you any idea what is the largest

number to show this phenomenon? You haven't? I'm
not surprised. The answer to the first question is 32043:

   $32043 \times 32043 = 1026753849$

The answer to the second is 99066:

   $99066 \times 99066 = 9814072356$

Once and once only
In these multiplication sums each of the digits from one
to nine is used just once:

   $4 \times 1738 = 6952$
   $4 \times 1963 = 7852$
   $12 \times 483 = 5796$
   $18 \times 297 = 5346$
   $27 \times 198 = 5346$
   $28 \times 157 = 4396$
   $39 \times 186 = 7254$
   $42 \times 138 = 5796$
   $48 \times 159 = 7632$

One to nine and just add two

   123456789
   987654321
   123456789
   987654321
   +   2
   ──────────
   2222222222

All the digits 45
Here are some simple sums that seem to lead
inexorably to the number 45.

   123456789 $= 45$ ($1+2+3+4+5+6+7$
                      $+8+9 = 45$)
   +123456789 $= 45$
   ─────────
   246913578 $= 45$ ($2+4+6+9+1+3+5$
                      $+7+8 = 45$)

$$987654321 \quad = 45 \quad (9+8+7+6+5+4+3$$
$$+2+1 = 45)$$
$$\underline{-123456789} \quad = 45$$
$$864197532 \quad = 45 \quad (8+6+4+1+9+7+5$$
$$+3+2 = 45)$$

$$123456789 \quad = 45$$
$$\times \quad\quad\quad 2$$
$$\overline{246913578} \quad = 45 \quad (2+4+6+9+1+3+5$$
$$+7+8 = 45)$$

$$987654321 \quad = 45$$
$$\times \quad\quad\quad 2$$
$$\overline{1975308642} \quad = 45 \quad (1+9+7+5+3+0+8$$
$$+6+4+2 = 45)$$

$$123456789 \quad = 45$$
$$\div \quad\quad\quad 2$$
$$\overline{61728394 \cdot 5} \quad = 45 \quad (6+1+7+2+8+3+9$$
$$+4+5 = 45)$$

$$987654321 \quad = 45$$
$$\div \quad\quad\quad 2$$
$$\overline{493827160 \cdot 5} \quad = 45 \quad (4+9+3+8+2+7+1$$
$$+6+0+5 = 45)$$

Now see what happens when you add the results of the two division sums:

$$617283945$$
$$\underline{+4938271605}$$
$$5555555550 \quad = 45 \quad (5+5+5+5+5+5+5$$
$$+5+5+0 = 45)$$

Now try adding the results of the two multiplication sums:

$$\begin{array}{r} 246913578 \\ +\,1975308642 \\ \hline 2222222220 \end{array} = 18\ (2{+}2{+}2{+}2{+}2{+}2{+}2$$
$$+2{+}2 = 18)$$

And: $1 + 8 = 9$

Finally add the two rows of digits in the sequences 1-9 and 9-1:

$$\begin{array}{r} 123456789 \\ +\,987654321 \\ \hline 1111111110 \end{array} = 9\ (1{+}1{+}1{+}1{+}1{+}1{+}1$$
$$+1{+}1 = 9)$$

and multiply this result by 5:

$$\begin{array}{r} 1111111110 \\ 5 \\ \hline 5555555550 \end{array} = 45\ (5{+}5{+}5{+}5{+}5{+}5{+}5$$
$$+5{+}5{+}0 = 45)$$

Two's company

$2 \times 5 = 10$ doesn't it? Well, by repeating the number two five times it's possible to produce the value of all the digits from zero to nine, or in this case from one to zero.

Watch:

$$2+2-2-2/2 = 1$$
$$2+2+2-2-2 = 2$$
$$2+2-2+2/2 = 3$$
$$2\times2\times2-2-2 = 4$$
$$2+2+2-2/2 = 5$$
$$2+2+2+2-2 = 6$$
$$22\div2-2-2 = 7$$
$$2\times2\times2+2-2 = 8$$
$$2\times2\times2+2/2 = 9$$
$$2-2/2-2/2 = 0$$

Blotting your copy-book

You'll recall the old excuse, 'I spilt the ink, sir'. It didn't cut much ice at school when you hadn't done your homework and it cuts even less ice now, because there is a better way of working out a sum like this one:

With the three figures left you can read, can you discover the missing ones, given that all of the ten digits from zero to nine were originally used in the calculation?

(Answer at the back.)

Sums of the century

I used to play Scrabble with a centenarian. He was called John Badley and he founded Bedales School in Hampshire where I was given (well, offered) my education. This is just the sort of mental recreation Mr Badley enjoyed in his eleventh decade: simple sums that only use the nine digits once and in their correct order and only require multiplication, addition or subtraction to come up with the same answer every time: 100.

$$1+2+3+4+5+6+7+(8\times9) = 100$$
$$-(1\times2)-3-4-5+(6\times7)+(8\times9) = 100$$
$$1+(2\times3)+(4\times5)-6+7+(8\times9) = 100$$
$$(1+2-3-4)(5-6-7-8-9) = 100$$
$$1+(2\times3)+4+5+67+8+9 = 100$$
$$(1\times2)+34+56+7-8+9 = 100$$
$$12+3-4+5+67+8+9 = 100$$
$$123-4-5-6-7+8-9 = 100$$

All but eight

```
        12345679
       ×99999999
        111111111
       111111111
      111111111
     111111111
    111111111
   111111111
  111111111
 111111111
1234567887654321
```

All the nines

All the digits from one to nine are used in these multiplication sums, all that is except for eight. However, if you multiply 12345679 by the first nine multiplies of nine you get some interesting answers, and if the digit in each product is multiplied by the number of digits in each product (nine) you'll find that another interesting product is produced.

$$12345679 \times 9 = 111111111$$
$$12345679 \times 18 = 222222222$$
$$12345679 \times 27 = 333333333$$
$$12345679 \times 36 = 444444444$$
$$12345679 \times 45 = 555555555$$
$$12345679 \times 54 = 666666666$$
$$12345679 \times 63 = 777777777$$
$$12345679 \times 72 = 888888888$$
$$12345679 \times 81 = 999999999$$

Digital squares and cubes

All the digits from one to nine appear in these two figures and their product, which happens to be the square of another number also using all the digits from one to nine:

$$246913578 \times 987654321 = 493827156^2$$

These two numbers and their square use the digits from one to nine:

$$854 \times 854 = 729316$$
$$567 \times 567 = 321489$$

Now try subtracting the cube of 641 from the cube of 642:

$$641 \times 641 \times 641 = 263374721$$

$$642 \times 642 \times 642 = 264609288$$

$$\begin{array}{r} 264609288 \\ -263374721 \\ \hline 1234567 \end{array}$$

SQUARE SURPRISES

Each of the numbers below uses three different digits. However, in each case you can add the three digits, square the result and then divide this into the original without leaving a remainder. These are the ten numbers:

162, 243, 324, 392, 405, 512, 605, 648, 810, 972

And this is what happens to each of them:

$$1 + 6 + 2 = 9$$
$$9 \times 9 = 81$$
$$162 \div 81 = 2$$

$$3 + 9 + 2 = 14$$
$$14 \times 14 = 196$$
$$392 \div 196 = 2$$

$$2 + 4 + 3 = 9$$
$$9 \times 9 = 81$$
$$243 \div 81 = 3$$

$$4 + 0 + 5 = 9$$
$$9 \times 9 = 81$$
$$405 \div 81 = 5$$

$$3 + 2 + 4 = 9$$
$$9 \times 9 = 81$$
$$324 \div 81 = 4$$

$$5 + 1 + 2 = 8$$
$$8 \times 8 = 64$$
$$512 \div 64 = 8$$

$$6 + 0 + 5 = 11$$
$$11 \times 11 = 121$$
$$605 \div 121 = 5$$

$$8 + 1 + 0 = 9$$
$$9 \times 9 = 81$$
$$810 \div 81 = 10$$

$$6 + 4 + 8 = 18$$
$$18 \times 18 = 324$$
$$648 \div 324 = 2$$

$$9 + 7 + 2 = 18$$
$$18 \times 18 = 324$$
$$972 \div 324 = 3$$

Now let us cast the net a little wider – to encompass any odd number in fact – and watch what we can do.

1 Choose any odd number and square it.
2 Find the two consecutive numbers which added together make this square.
3 Square these two numbers.
4 The difference between these two squares will equal the square of the original odd number!

Here are a couple of examples:

1 $\qquad 19^2 = 361$
2 $\quad 180 + 181 = 361$
3 $\qquad 181^2 = 32761$
$\qquad 180^2 = \underline{32400}$
4 Subtract $\quad = \quad 361$

1 $\qquad 97^2 = 9409$
2 $\quad 4704 + 4705 = 9409$
3 $\qquad 4705^2 = 22137025$
$\qquad 4704^2 = \underline{22127616}$
4 Subtract $\quad = \quad 9409$

CUBE CURIOS

152, 251 and 237 may not strike you immediately as sharing much in common as numbers, but they can be made to produce the cubes of three, four and five respectively, if you follow this set of rules:

1 Square any of these three numbers.
2 Add the first two and the last two digits of each square.
3 And you'll find that the answer will equal the cube of either three, four or five depending on which number you use.

I'll show you:

      152
1    $152^2 = 23104$
2 $23 + 04 = 27$
3      $3^3 = 27$

      251
1    $251^2 = 63001$
2 $63 + 01 = 64$
3      $4^3 = 64$

      237
1    $237^2 = 56169$
2 $56 + 69 = 125$
3      $5^3 = 125$

Make me a number
Now it's your turn. I'd like you to think up a number that uses some of the digits from one to nine once only, and to think up another number which uses the remainder of the digits, *and* which is double the first number. Got it? Good.
(Answer at the back.)

In this second (much easier) problem, all you've got to do is think of a way of writing one by using all the digits from zero to nine.
(Answer at the back.)

Paired products

There's more work for you here, I'm afraid. Kindly write down any three-digit number. Now multiply it by 11, and then multiply the product by 91. Have a look at your total and you'll see that you've actually produced your original number twice. Try doing it again with a different three-digit number and watch it happen all over again. Look:

$$
\begin{array}{r}
222 \\
\times 11 \\
\hline
2442 \\
\times 91 \\
\hline
222222
\end{array}
\qquad \text{and:} \qquad
\begin{array}{r}
719 \\
\times 11 \\
\hline
7909 \\
\times 91 \\
\hline
719719
\end{array}
$$

ENQUIRE WITHIN

THREE IN THREE

Within the diagram on page 43 you should be able to find three THREEs – written horizontally or vertically or diagonally, forwards or backwards. (This chapter shouldn't be called Enquire Within. It should really be called Enquire at the Back, because that's where you will find all the answers.)

ONE IN THREE

These days one in three marriages ends in divorce. Apparently the critical years in any marriage are those just before your Tin and just after your China anniversaries. If you're hoping to make it to your Golden anniversary, you will want – and need – to know what the anniversaries on the way are called. This quiz should help you. All you have to do is look at the list of objects and materials and decide with which anniversaries they are associated. (If you don't know the answers, enquire first at the back and then at your local Marriage Guidance Council!)

The objects and materials
1 Platinum
2 Wool, copper
3 Ivory
4 Pearl
5 Ruby
6 Pottery, willow
7 Emerald
8 Sapphire
9 Cotton

The objects and materials

10 Fruit, flowers	**16** Lace	**22** Wooden
11 Bronze, pottery	**17** Paper	**23** China
12 Tin	**18** Crystal	**24** Golden
13 Steel	**19** Sugar	**25** Diamond
14 Silver	**20** Leather	
15 Coral	**21** Silk, linen	

The anniversaries

First	Tenth	Twenty-fifth
Second	Eleventh	Thirtieth
Third	Twelfth	Thirty-fifth
Fourth	Thirteenth	Fortieth
Fifth	Fourteenth	Forty-fifth
Sixth	Fifteenth	Fiftieth
Seventh	Twentieth	Fifty-fifth
Eight		Sixtieth
Ninth		Seventieth

MOVIE NUMBERS

Numbers have featured within the titles of many films – *Five Easy Pieces*, *Snow White and the Seven Dwarfs*, *The House on 92ND Street* were three of my favourites – but how much do you remember about them? This quiz should show you.
(To check your answers, enquire at the back.)

1. What was the title of Ray Bradbury's novel, later filmed by Francois Truffaut, that was all about burning books?
2. It won an Academy Award in 1963 as the best foreign film. It was the work by an eminent Italian director. What was it called?
3. Dickens' great novel about the French Revolution has been filmed four times. Who played the lead in the 1958 version?

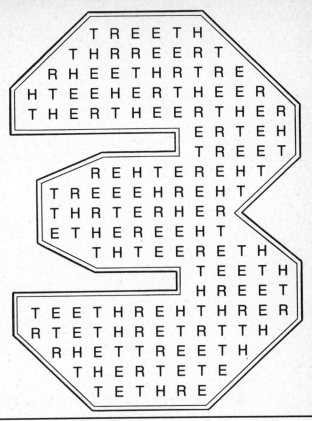

```
        T R E E T H
      T H R R E E R T
    R H E E T H R T R E
  H T E E H E R T H E E R
  T H E R T H E E R T H E R
              E R T E H
              T R E E T
      R E H T E R E H T
    T R E E E H R E H T
    T H R T E R H E R
    E T H E R E E H T
      T H T E E R E T H
                T E E T H
                H R E E T
    T E E T H R E H T H R E R
    R T E T H R E T R T T H
    R H E T T R E E T H
      T H E R T E T E
      T E T H R E
```

4 Who composed the music for *Henry V*, filmed in 1944?

5 One of the most famous films to come out of Japan lent its theme and part of its title to a well known western. What were the two films called?

6 Billy Wilder shot this film in 1943. Rommel is the chief character. What was it called?

7 The plot deals with a submarine full of Nazis on the run in Canada, led by Eric Portman. Can you name the film?

8 Cecil B de Mille cast Charlton Heston as Moses in the 1956 version of *The Ten Commandments*.

Who played the Pharaoh?

9   Ingrid Bergman played Gladys Aylward and it was Robert Donat's last film. What was it called?

10 First there were three and in the sequel there were *The Four Musketeers*. Can you name the actors who played them in the 1973 film?

11 Who wrote the script for the famous 1949 film of *The Third Man*?

12 The film was made in 1966 and first brought Raquel Welch to public notice, with reason. What was the film's title?

13 Who wrote the novel that Stanley Kubrick filmed in 1968 with spellbinding, if obscure, effect?

14 In 1939 Alexander Korda filmed this famous story with John Clements and Ralph Richardson in the leads. The setting was the Sudan. What was the film?

15 Robert Donat played Richard Hannay in the 1935 version of Buchan's novel, Robert Powell played him in the most recent version. But which British actor came in between them, in the 1959 version of *The Thirty-Nine Steps*?

16 This film, starring Marlon Brando, was initially banned in almost every cinema in the UK and was only given a certificate in 1967, 13 years after it first appeared. What was it called?

17 How long did David Niven and the others spend in the Chinese capital?

18 Gregory Peck finally cracked up under the strain. The film was made several years after the events in question, but it was still up with the times. What was it called?

19 Who wrote the original story in which a much travelled David Niven starred in the 1956 award-winning film version?

20 It's usually regarded as Bergman's most fascinating film. Which Bergman is it and which film?

FANTASTIC FIGURES

The fantastic figures I've got in mind don't belong to Marie Antoinette or Jayne Mansfield – who shared identical bust measurements by the way. Indeed, the figures aren't even numbers. They're a group of individuals I call the Famous Five. . .people from the past who simply had a phenomenal way with mental arithmetic. Some of their feats may not seem so fantastic in the age of the universal pocket calculator, but in their way and in their time each of them was something of a genius.

JEDEDIAH BUXTON *(c.1707-1772)*

As his surname suggests, Buxton was a native of Derbyshire. He was the son of the village schoolmaster in Elmton, but in spite of this apparent advantage, he never learned to write or perform arithmetic on paper. Buxton was probably the least spectacular of the Famous Five, but even his 'modest' powers would leave most of us standing.

Among the mental challenges Buxton rose to were these:

. . .How many acres are there in a rectangular field 351 yards long by 261 wide? He gave the answer in 11 minutes.

45

. . .If sound travels 1,142 feet in a second, how long will it take to travel five miles? He answered this in 15 minutes.

. . .Asked the sum to which a farthing would increase if it were doubled 140 times, Buxton eventually produced the answer that the number of pounds requires 39 digits to express it, with two shillings and eight pence left as remainders.

Buxton was then asked to square this number of 39 digits. He calculated the answer to this problem at intervals over the next two and a half months – and all in his head.

ZERAH COLBURN *(1804-1840)*

In his short life Zerah Colburn established himself as one of the most remarkable exponents of mental arithmetic on the stage in both America and England. He was touring the USA at the age of six and made his London debut two years later. But as he grew older – after he was educated ironically – he showed less of his youthful promise. Having tried his luck on the stage, as a schoolmaster, as a preacher and finally as a language teacher, he died at the age of 36, after writing an autobiography in which he outlined his skills.

. . .Asked to raise eight to its sixteenth power, Colburn came back with the correct answer. 281,474,976,710,656, within a matter of seconds.

. . .When he was asked to raise the digits from two to nine each to its tenth power, he rattled off the answers at such a speed that the man recording them had to ask him to slow down and repeat what he'd said.

. . .Factors were his forte. When he was asked to give the factors of 268,336,125 he replied 941 and 263. The factors of 171,395 came back almost as quickly as five, seven 59 and 83. And when challenged to give the factors for 36,083, he replied that there were none. He once admitted that he would factorize a number whenever it was convenient, so when asked to multiply 21,734 by 543, he produced the answer, 11,801,562, by multiplying 65,202 by 181.

JOHANN MARTIN ZACHARIAS DASE (1824-1861)

Dase was born in Hamburg and unlike other calculating prodigies received a fair education. He made a number of contributions to science during his short life, but is best remembered for the 'calculating exhibitions' given in Germany, Austria and England. Among his recorded feats were:

. . . Multiplying 79,532,853 by 93,758,479, which he answered in 54 seconds.

. . .Finding the product of two 20-digit numbers, which took him six minutes.

. . .Finding the product of two 40-digit numbers, which took him 40 minutes.

. . .Finding the product of two 100-digit numbers, which took him eight hours 45 minutes.

. . .After looking at a set of dominoes for a second, he was able to give the sum of their dots, 117. He could tell you how many letters there were in any line of print selected at random, and shown a row of 12 digits he would be able to memorise them in half a second to the extent of knowing what they were and where in the row they belonged.

TRUMAN HENRY SAFFORD (1836-1901)

Safford was unusual among child prodigies of mental arithmetic in that his powers were not totally blunted by education. He went on to graduate from Harvard and turned his skills to astronomy, although his mathematical powers in adult life were never as keen as they had been when he was a child. After amusing and entertaining his family when he was young, he was examined for the first time when he was ten years old, and these were some of the questions and answers that confirmed his genius:

. . .Extracting the cube from a number of seven digits, which he was able to do instantly.

. . .Asked the surface of a regular pyramid with a slant height of 17 feet, and pentagonal base with each side measuring 33.5 feet, Safford pondered for two minutes and then announced 3354.5558 square feet.

GEORGE PARKER BIDDER
(1806-1878)

Bidder was a mathematical genius who managed to retain most of his powers until his death and, having had a formal education, was able to assess and summarise his gift.

His origins were humble enough. He was the son of a Devon stonemason. A brief spell at the village school left him with little arithmetical ability, but by learning to count to 100 at the age of six, he gradually taught himself the principles of addition, multiplication and subtraction, by arranging and rearranging objects like buttons or marbles. At the age of nine, he starting touring the country to show off his skills, and earn his father a bob or two. During these tours, he was frequently pitted against Colburn, whom he usually outshone. These are just a few of the staggering feats he performed while touring between 1815 and 1819:

. . .When he was nine he was asked how long it would take the inhabitants of the moon the hear the battle of Waterloo, if sound travelled at four miles a second, and the moon was 123,256 miles from the earth. In less than a minute Bidder came back with the answer 21 days, nine hours and 34 minutes.

. . .When he was ten and, though able to write, unable to write numbers, he was asked what would be the interest on £11,111 for 11,111 days at five per cent per annum. It took young Bidder a minute to reply £16,911. 11 shillings.

. . .It took him just 35 seconds to tell one questioner how many hogsheads of cider could be made from a million apples, if 30 apples made one quart. The answer was 132 hogsheads, 17 gallons, one quart and ten apples over. (By the way, a hogshead holds 54 gallons of beer and cider can't be very different.)

. . .The famous astronomer, Sir William Herschel, once asked Bidder how far a certain star was from the earth, given that light travels from the sun to the earth in eight minutes, and the sun is 98,000,000 miles away, when it takes six years four months for light from the star to reach earth, working on the basis of each year containing 365 days six hours and 28 days to a month. Bidder told him the star would be 40,633,740,000,000 miles away.

. . .In one minute he worked out the value in pounds of a ring of pennies circling the earth when each penny had a diameter of $1.\frac{3}{8}$ inches. The answer was £4,803,340.

. . .When he was 14 he was asked to give a number which had a cube, that less 10 multiplied by its cube, was equal to the cube of six. Instantly Bidder gave the correct answer three.

What makes Bidder's ability all the more interesting from our point of view is his own assessment in later life. Curiously he asserted that he was *not* gifted with a particularly retentive memory, despite the fact that he once read a number backwards and immediately repeated it the right way round, and what's more repeated it correctly an hour later. The number was: 2,563,721,987,653,461,598,746,231,905,607,541,128,

975,231!

Probably his most amazing feats were those involving multiplication. He explained his system for this. Supposing he wanted to multiply 397 by 173, he would have set about it in this way:

Starting with	$100 \times 397 = 39700$	
Add	$70 \times 300 = 21000$	making 60,700
Add	$70 \times\ \ 90 =\ \ 6300$	making 67,000
Add	$70 \times\ \ \ \ 7 =\ \ \ \ 490$	making 67,490
Add	$3 \times 300 =\ \ \ \ 900$	making 68,390
Add	$3 \times\ \ 90 =\ \ \ \ 270$	making 68,660
Add	$3 \times\ \ \ \ 7 =\ \ \ \ \ \ 21$	making 68,681

That may seem fairly easy when set out as it is, but we have to remember that Bidder would perform that calculation so fast that it would appear to be almost instantaneous. This gave him the use of what amounted to a multiplication table that went up to 1000 × 1000, so that when tackling very large numbers of, say nine digits, he was able to deal with them in units of three, if they were in a scale with a radix of 1000. As a result he could calculate a multiplication of two nine-digit numbers in about six minutes.

FEATS OF MEMORY

I'm afraid I can't turn you into a mathematical phenomenon to rival one of the Famous Five, but I can have a go at helping you to remember numbers you may need or want to remember by introducing you to the Number Alphabet. It was originally devised by a Frenchman named Pierre Hérigone in the mid-seventeenth century. It's been modified and improved since then and is now widely used by memory experts throughout the world.

As you'll see, the system works on the principle of

substituting consonants for the digits from one to zero. Once the relevant consonants have been chosen all you have to do is to choose vowels that will transform them into a word, and you'll be able to memorise the word to remind you of the number. What the Number Alphabet does in fact is help you form personal mnemonics. Of course, the first thing you've got to do is remember the alphabet.

Digits	Consonants	Memory aids
1	T, D	T has *1* downward stroke
2	N	N has *2* downward strokes
3	M	M has *3* downward strokes
4	R	R is the *4*th letter in 4
5	L	L is *50* in Roman numerals
6	J, G, SH, CH	J looks like *6* when it is reversed
7	K, hard G hard C	K can be printed with two *7*'s
8	F, V, PH as in photo	F looks like *8* in lower case script
9	P, B	P looks like *9* when it is reversed
0	Z, S, C	Z is the first letter of zero, *0*

GOES INTO…

WILL IT GO?

The next time you're faced with a larger number that you want to divide by either seven, 11 or 13, save yourself a little trouble by checking quickly whether your divisor will go into the number without leaving any remainders. This is how you do it:

For division by seven
a) Write down the number you want to divide.
b) Beginning with the last digit, group the number in sets of three digits with a plus sign and a minus sign alternating between the sets to form a sum.
c) Work through this sum.
d) Look at your answer and see if this can be divided evenly by seven. If it can, then the original number will also be divisible by seven, without leaving a remainder.

Watch:
a) Your number is, say: 3722544
b) Grouped in sets it is: $3 - 722 + 544$
c) This equals: $- 175$
d) Divided by 7 this equals $- 25$. So 3722544 can be evenly divided by 7. In fact it equals 531792.

For division by eleven
a) This time you write down the number you want to

divide, one digit at a time, placing plus and minus signs between the digits alternately.
b) Then you work through the sum as before.
c) If your answer is divisible by 11, then you'll know that that number can be divided by 11 without leaving a remainder, e.g. $9-1+8-0+6=22$

Watch:
a) Your number is, say, 9151494, which written out as a sum is: $9 - 1 + 5 - 1 + 4 - 9 + 4$.
b) Work through this.
c) The answer is 11.
d) So you know that 9151494 can be divided evenly by 11.
It equals 831954 in fact.

For division by thirteen
a) This method is similar to the one used for seven. So write out your number in sets of three, starting as the end and working backwards with alternating plus and minus signs.
b) Work through this sum.
c) If your answer can be divided by 13, then you know that the larger number can be evenly divided by 13 also.

Watch:
a) Your number is, say, 5618197
b) Grouped in sets it is: $5 - 618 + 197$
c) Work through this and it equals $- 416$
d) Divide this by 13 and it equals $- 32$. So 5618197 can be divided evenly by 13. It actually goes in to it 432169 times.

You can use seven, 11 and 13 in an entertaining division trick too. All you need is a three-digit number, anything from 100 to 999. You then make this a six digit-number by repeating the three digits in the same order and divide it first by seven, then by 11 and finally by 13. And the result? You'll end up with the number you began with. This is how it works:

a) Start with 456.
b) Make it a six-digit number 456456.
c) Divide by 7 to make 65208.
d) Divide this by 11 to make 5928.
e) Divide by 13 to make 456, the number you began with.

Try it again:
a) Start with 913.
b) Make it a six-digit number 913913.
c) Divide by 7 to make 130559.
d) Divide by 11 to make 11869.
e) Divide by 13 to make 913, the number you began with.

NINE INTO ANYTHING WILL GO

Well, that's almost true. It will go, provided you start with a number that has more than one digit (in other words anything above ten) and follow these simple steps:

a) Write down your number.
b) Reverse the digits
c) Subtract the lower number from the upper.
d) And you'll find that your answer can be divided by nine without leaving a remainder.

This rule applies from ten to 10 billion, but in order to conserve space and nervous energy I'm going to give relatively modest figures for my examples.

$$\begin{array}{r} 12 \\ -\ 21 \\ \hline 9 \end{array}$$ and 9 divided by 9 equals 1

```
  6493
- 3946
  2547 and 9 goes into that 283 times

  62481753
- 35718426
  26763327 and 9 goes into that 2973703 times.
```

EIGHT-DIGIT DIVISION

One of the more amusing aspects of division concerns the four-digit number written twice to make an eight-digit number. This can always be evenly divided by both 73 and 137, neither of which look very suitable at first sight. However, here's proof:

$$\frac{1031747}{73/75317531}$$

and

$$\frac{549763}{137/75317531}$$

or

$$\frac{1353012}{73/98769876}$$

and

$$\frac{720948}{137/98769876}$$

As a bonus, you'll also find that if the final digit of your number is zero or five, the number will also be divisible by 365 and 50005. Look:

$$\frac{353460}{73/25802580}$$

and

$$\frac{188340}{137/25802580}$$

and

$$\frac{70692}{365/25802580}$$

and
$$\frac{516}{50005/25802580}$$

2519 DIVIDED

Divide 2519 by the numbers from one to ten and watch the pattern of remainders:

$$\frac{2519}{1/2519}$$ remainder 0

$$\frac{1259}{2/2519}$$ remainder 1

$$\frac{839}{3/2519}$$ remainder 2

$$\frac{629}{4/2519}$$ remainder 3

$$\frac{503}{5/2519}$$ remainder 4

$$\frac{419}{6/2519}$$ remainder 5

$$\frac{359}{7/2519}$$ remainder 6

$$\frac{314}{8/2519}$$ remainder 7

$$\frac{279}{9/2519}$$ remainder 8

$$\frac{251}{10/2519}$$ remainder 9

DIVIDE AND RULE

Study the circle on page 57. The 11 dots represent two

shapes in the circle. The eight on the outside form a square, the three in the middle form a triangle. That's obvious enough. What isn't obvious, however, is how you draw straight lines across the circle to divide it into 11 areas, each with only one dot inside it, and no area without a dot. Can you divide the circle in this way? (Answer at the back.)

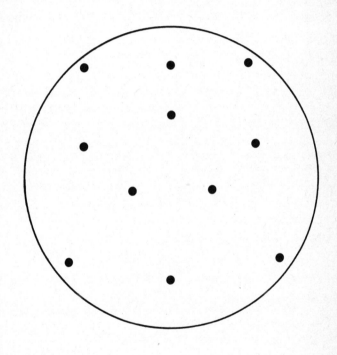

HISTORICAL CONTEXTS

A ROYAL LINE

You might be forgiven for blanching when asked to remember all the kings and queens of England back to William I, especially when there are eight Edwards, six Georges and eight Henries to contend with. To help you and conveniently forgetting what happened before Duke William landed in 1066, here's a traditional mnemonic that will give them to you in their correct order and with their post-nominal numbers. Starting with William I, it goes:

Willie, Willie, Henry, Stee,
Henry, Dick, John, Henry Three,
One, two, three Neds, Richard Two,
Henry Four, Five, Six, then who?
Edward Four, Five, Dick the Bad,
Henries twain and Ned the Lad,
Mary, Bessie, James the Vain,
Charlie, Charlie, James again,
William and Mary, Anna Gloria,
Four Georges, William, and Victoria;
Edward Seven came next and then
George the Fifth in 1910.
Edward the Eighth soon abdicated

And so a George was reinstated.
Finally, we must draw breath
And give three cheers for Elizabeth!

NUMBER ME A KING

When considering the British royal line, it's interesting to note that the three kings who have each reigned for over 50 years, have all been the third of their name:
Henry III, reigned for 56 years, until 1272.
Edward III was king for 50 years, until 1377.
And George III reigned for 59 years, until 1820.

ALL IN THE FAMILY

In case you've ever puzzled over the way in which the British royal line is inter-related and how all those post-nominal numbers finally produced Queen Elizabeth II, allow me to give you a simple outline of what went on:
QUEEN ELIZABETH II is the daughter of . . .
GEORGE VI, who was the brother of . . .
EDWARD VIII, who was the son of . . .
GEORGE V, who was the son of . . .
EDWARD VII, who was the son of . . .
QUEEN VICTORIA, who was the niece of . . .
WILLIAM IV, who was the brother of . . .
GEORGE IV, who was the son of . . .
GEORGE III, who was the grandson of . . .
GEORGE II, who was the son of . . .
GEORGE I, who was the cousin of . . .
ANNE, who was the sister-in-law of . . .
WILLIAM III, who was the son-in-law of . . .
JAMES II, who was the brother of . . .
CHARLES II, who was the son of . . .
CHARLES I, who was the son of . . .

JAMES I, who was the cousin of . . .
ELIZABETH I, who was the half-sister of . . .
MARY I, who was the half-sister of . . .
EDWARD VI, who was the son of . . .
HENRY VIII, who was the son of . . .
HENRY VII, who was the cousin of . . .
RICHARD III, who was the uncle of . . .
EDWARD V, who was the son of . . .
EDWARD IV, who was the cousin of . . .
HENRY VI, who was the son of . . .
HENRY V, who was the son of . . .
HENRY IV, who was the cousin of . . .
RICHARD II, who was the grandson of . . .
EDWARD III, who was the son of . . .
EDWARD II, who was the son of . . .
EDWARD I, who was the son of . . .
HENRY III, who was the son of . . .
JOHN, who was the brother of . . .
RICHARD I, who was the son of . . .
HENRY II, who was the cousin of . . .
STEPHEN, who was the cousin of . . .
HENRY I, who was the brother of . . .
WILLIAM II, who was the son of . . .
WILLIAM I . . .

And I think we can leave it there, don't you? I'm sure
you know the Queen is King Alfred the Great's
granddaughter 36 times removed anyway.

IN DAYS OF YORE

Complaints are commonly aired today about contem-
porary self-indulgence and the grossness of our age.
However, in terms of food and drink at least, our
forefathers could certainly have taught us a thing or
two. Take a mouth-watering look at this catalogue of
goodies. It is just *part* of the inventory of comestibles

consumed at the feast celebrating the installation of the Bishop of York in 1464:

... 300 quarters of wheat
... 300 tuns of ale
... 100 tuns of wine
... 1 pipe of hippocras
... 104 oxen
... 6 wild bulls
... 1000 sheep
... 304 calves
... 304 'porkes'
... 400 swans
... 2000 geese
... 1000 capons
... 2000 pigs
... 104 peacocks
... 13500 birds of other sorts
... 500 + stags, bucks and roes
... 1500 hot venison pasties
... 608 pikes and breams
... 12 porpoises and seals
... 1300 dishes of jelly, cold baked tarts, hot and
     cold custards and 'spices sugared delicates, and
     wafers plentie' . . .

And all this was required for a mere 600 official guests!

Parliamentary elections haven't always been such dull affairs either, Even as recently as the early years of the nineteenth century a voter could be sure of getting more from his candidate than a few well-rehearsed platitudes about the economy and the local by-pass. Once upon a time you got real value for your vote. This is the candidate's bill from just one small pot-house in Ilchester, Somerset, showing the cost of one day on the hustings – such a good day in fact that the calculation leaves something to be desired!

353 bottles rum and gin at 6s	£105	18	0
57 bottles French brandy at 10s 6d	29	18	6
514 gallons of beer at 2s 8d	68	18	8
792 dinners at 2s 6d	99	0	0
	£304	17	2

Cheers!

EIGHT OF THE BEST

What do Euclid, Pascal, Bertrand Russell, Mae West, Carl Sanburg, Sydney Smith, Alfred North Whitehead and Francis Bacon have in common?

They each had something special to say about the world of numbers. The question is: who said what?

1 If a man's wit be wandering, let him study the mathematics.

2 There is no royal road to geometry.

3 Mathematicians who are only mathematicians have exact minds, provided all things are explained to them by means of definitions and axioms; otherwise they are inaccurate and insufferable, for they are only right when the principles are quite clear.

4 Mathematics may be defined as the subject in which we never know what we are talking about, nor whether what we are saying is true.

5 Arithmetic is where the answer is right and everything is nice and you can look out of the window and see the blue sky – or the answer is wrong and you have to start all over and try again and see how it comes out this time.

6 What would life be without arithmetic, but a scene of horrors?

7 A man has one hundred dollars and you leave him with two dollars, that's subtraction.

8 Mathematics is thought moving in the sphere of complete abstraction from any particular instance of what it is talking about.

I T'S A SQUARE WORLD

THE MAGIC OF THE SQUARES

In the world of numbers 'magic squares' consist of sets of figures arranged in a square shape, so that the sum of the numbers in every row, every column and every diagonal is the same.

Magic squares have been in existence for thousands of years. In ancient China and other early civilizations, they were believed to have mystical powers, often in association with the powers attributed to certain specific numbers. The Chinese classic book of divination, the *I Ching*, contains this magic square of nine numbers, which add up to 15 in the rows, columns and diagonals:

8	1	6
3	5	7
4	9	2

In the Middle Ages magic squares were sometimes engraved on silver plates as charms to ward off the plague. Probably the most famous example of this practice is the magic charm that appears on just such a plate in Albrecht Dürer's picture of *Melancholy*, which was engraved in 1514, as the middle numbers of the bottom line reveal:

16	3	2	13
5	10	11	8
9	6	7	12
4	15	14	1

Following Dürer's example, here are two more 4 × 4 magic squares. This one has columns, rows and diagonals that add up to 62:

23	10	9	20
12	17	18	15
16	13	14	19
11	22	21	8

And this one has columns, rows and diagonals that add up to 150:

45	32	32	42
34	39	40	37
38	35	36	41
33	44	43	30

Now take a look at this four-figure magic square. The same rules of addition apply, but the numbers are considerably larger.

2243	1341	3142
3141	2242	1343
1342	3143	2241

This next square is even more magical than most. Only the numbers one and eight are used. The total of every row, column and diagonal is 19,998. The four corners also add up to this figure. What's more, if you turn the page upside down and work through the magic square that way, you'll see that it still produces the same sums when you add the numbers in the rows, columns, diagonals and corners. Finally, find a mirror, hold the magic square up in front of it, and perform the same additions. You'll see that it still works!

8818	1111	8188	1881
8181	1888	8811	1118
1811	8118	1181	8888
1188	8881	1818	8111

Here are a couple of unusual 5 × 5 magic squares. In the first all the numbers between one and 25 are used just once, giving a row/column/diagonal total of 65:

17	24	1	8	15
23	5	7	14	16
4	6	13	20	22
10	12	19	21	3
11	18	25	2	9

Look at this one and you'll see that as well as the rows, columns and diagonals adding up to 105, the four corners plus the number in the centre add up to 105 as well:

25	32	9	16	23
31	13	15	22	24
12	14	21	28	30
18	20	27	29	11
19	26	33	10	17

There is no limit to the size of a magic square. This monster, for example, contains 10 × 10 numbers, with rows, columns and diagonals that add up to 505:

92	99	1	8	15	67	74	51	58	40
98	80	7	14	16	73	55	57	64	41
4	81	88	20	22	54	56	63	70	47
85	87	19	21	3	60	62	69	71	28
86	93	25	2	9	61	68	75	52	34
17	24	76	83	90	42	49	26	33	65
23	5	82	89	91	48	30	32	39	66
79	6	13	95	97	29	31	38	45	72
10	12	94	96	78	35	37	44	46	53
11	18	100	77	84	36	43	50	27	59

As you've no doubt realised, every number between one and 100 is in the square somewhere.

Here is a magic square in which you don't add the numbers: you multiply them and arrive at a total of 4096 in each line:

128	1	32
4	16	64
8	256	2

In this next square you can either add or multiply the numbers. Adding the rows, columns and diagonals gives a total of 840. Multiplying gives a total of 2,058,068,231,856,000!

46	81	117	102	15	76	200	203
19	60	232	175	54	69	153	78
216	161	17	52	171	90	58	75
135	114	50	87	184	189	13	68
150	261	45	38	91	136	92	27
119	104	108	23	174	225	57	30
116	25	133	120	51	26	162	207
39	34	138	243	100	29	105	152

Three tricky squares

Try adding these three magic squares and you'll run into trouble. The total of the rows, columns and diagonals refuse to agree. All three have a common total for the rows, columns and diagonals, but to achieve it, you may have to subtract as well as add.

2	1	6		2	1	4		8	1	4
3	5	7		3	5	7		3	5	7
4	9	8		6	9	8		6	9	2

(Answers at the back.)

Do-It-Yourself
To construct a 4 × 4 magic square of your own, choose any sequence of numbers, put the lowest in the bottom right-hand square, then follow the order of numbering set out below.

16th	3rd	2nd	13th
5th	10th	11th	8th
9th	6th	7th	12th
4th	15th	14th	1st

The pattern for a 5 × 5 square is this:

17th	24th	1st	8th	15th
23rd	5th	7th	14th	16th
4th	6th	13th	20th	22nd
10th	12th	19th	21st	3rd
11th	18th	25th	2nd	9th

The square to end all magic squares
This magic square provides you with a mesmerising trick designed to leave any audience flabbergasted.

Here is the square:

125	191	248	169	116
48	114	171	92	39
136	202	259	180	127
69	135	192	113	60
64	130	187	108	55

And this is how you use it:

a) Ask your victim to select one of the numbers in the square.

b) Ask him or her to cross off all the numbers that are vertically or horizontally in line with that one.

c) Now ask him or her to repeat this process four times, eliminating as he or she goes, until you are left with no more numbers to eliminate. There should be five uncrossed numbers left.

d) But before your victim starts adding them up, announce that the sum of these numbers is 666. It always is, so you could even perform the trick with your eyes closed!

Here's how it went the last time I did it:

125	191	248	169	116
48	114	171	92	39
136	202	259	180	127
69	135	192	113	60
64	130	187	108	55

$$125 + 92 + 202 + 60 + 187 = 666$$

Now you have a go.

Domino Magic Squares

A normal set of dominoes contains 28 dominoes with values ranging from double nought to double six. You can arrange all 28 dominoes to form a 7 × 7 square, with a column of blanks that you disregard. The pips in each row, column and diagonal of the square total 24:

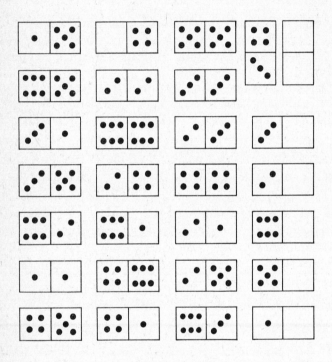

By removing some of the dominoes, it's possible to form other squares. This one has rows, columns and diagonals that add up to 19:

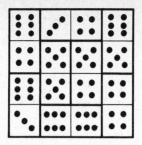

Here the rows, columns and diagonals add up to 13:

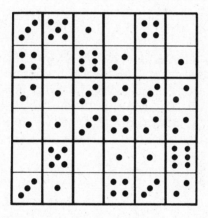

J ARGON - THE LANGUAGE OF NUMBERS

WORDS YOU CAN COUNT ON

Rummaging through a variety of English dictionaries loaned to me by my friend the numerate wordaholic Darryl Francis, I discovered that **SEVENTY-FOUR**, as well as being 74, turns out to be a South African fish. What's more, **SIXTY-SIX** is a two-handed card game, **FORTY-NINE** is a lunch counter customer who doesn't pay.

It seems that quite a number of numerals lead double-lives. Here are three dozen to be starting with. Some numbers can have more than one meaning, but I have simply listed my favourites.

ZERO	a place in Lauderdale County, Mississippi
ONE	the ultimate being
TWO	a two-dollar bill
THREE	a rugby three-quarter
FOUR	a type of racing boat
FIVE	a basketball team
SIX	a chemical, properly called arsphenamine

SEVEN	the oarsman immediately behind the stroke in an . . .
EIGHT	the boat in which they row
NINE	a baseball team
TEN	a measure of coal, from 48 to 50 tons
ELEVEN	a football team
TWELVE	a shilling
THIRTEEN	an Irish term for an English shilling
FOURTEEN	a special order
FIFTEEN	the first score in a game of tennis
SIXTEEN	a place in Meagher County, Montana
SEVENTEEN	a corpse
EIGHTEEN	the size of a piece of paper cut eighteen from a sheet
NINETEEN	the score of zero in cribbage
TWENTY	a twenty-dollar bill
TWENTY-ONE	limeade
TWENTY-TWO	rifle or pistol with a ·22 calibre
TWENTY-THREE	the end
TWENTY-FOUR	a day
TWENTY-FIVE	a variety of spoil-five
TWENTY-SIX	a gambling game involving the use of dice
TWENTY-SEVEN	three 'nine' cards
TWENTY-EIGHT	a Western Australian parakeet
TWENTY-NINE	an article of clothing in the 29th size
THIRTY	a mark or sign of completion
THIRTY-ONE	any of various card games
THIRTY-TWO	a pistol with a ·32 calibre
THIRTY-THREE	an order of ground beef steak
THIRTY-FOUR	go away!

| THIRTY-FIVE | a gambling game for from two to five players |
| THIRTY-SIX | a spotlight with a 36-inch lens |

CB SPEAK

CB fanatics all over the English-speaking world have a language of their own and it is a language in which numbers play a significant part. Here are some of the CB numbers that have specific meanings. I am offering them to you in case you're ever travelling in a vehicle equipped with a CB radio and suddenly find yourself in need of anything from a 'location' (Twenty) to a 'lavatory' (ten-hundred room):

Big eights	best wishes; sign off
Big ten-four	OK!
Big threes	best wishes; sign off
Break Channel 21 etc	breaking into specified channel
Break 21 etc.	breaking into particular channel
Double eighteen	two juggernauts driving side by side
Double eighty-eights	love and kisses
Double seven	no; no contact
807	beer
Eighteen wheeler	large lorry; juggernaut
Eights	best wishes
Eights and other good numbers	best wishes
Eighty-eights	love and kisses
Eighty-eights around the house	best wishes to you and your family
Five by nine	clear, strong radio signal
Five-finger discount	stolen goods
Five-two	(half of ten/four); perhaps

Fiver	short conversation
Forty-footer	juggernaut
Forty-fours	love and kisses
Forty roger	OK
Four	abbreviation of ten/four
Four-legged go-go dancers	pigs
Four roger	yes
Four ten	ten/four!
Four-wheeled frog	French car
Four-wheeled log	skateboard
Four-wheeler	car
Four-wheeler fever	car trouble
Gimme five	speak to me for a few minutes
Go break 21 etc	permission to break into specified channel
Go ten-one hundred	stop to go to the lavatory
Gone permanently ten-seven	dead
Good numbers	best wishes
Good pair	best wishes
Got a ten-two	getting clear reception
Have a 36-24-36 tonight	have a good evening
Hit me one time	answer me back so that I can check my radio
Home twenty	home address
Hundred mile coffee	strong coffee
Making three tracks in the sand	very tired
Meeting twenty	meeting place
Nine-to-fivers	people who work a regular day with regular hours
Niner	emergency
On the seventy	driving at 70 mph
One foot on the floor one hanging	

out the door	driving at full speed
One time	immediately
Other half	driving partner
School twenty	location of school
Seven threes	best wishes
Seventeen wheeler	juggernaut with a puncture
Seventy-three	best wishes
Six-ten	make love
Six-wheeler	small lorry; passenger car with trailer
Six and eights	best wishes
Smokey of four legs	mounted police
Smokey two-wheeler	police motor-cyclist
Squeeze-n-eights	love and kisses
Stack them eights	best wishes
Ten bye-bye	goodbye
Ten-hundred room	lavatory
Ten-roger	OK
Ten-ten and listin' in	monitoring without transmitting
Ten-ten 'till we do it again	goodbye
Thirty-three	emergency
Thirty-twelve	(ten/four three times) very definitely
Three lane dog pound	crowded road
Three-legged four-wheeler	broken-down car
Threes	general salutation
Threes and eights	best wishes
Three on you	best wishes
Turn twenty	turn
Twenty	location
Two-stool beaver	fat woman
Two-way strip	dual carriageway

Two-wheeler	motorcycle
U-ville 1	Oxford
U-ville 2	Cambridge
What's your eighteen?	what kind of lorry are you driving?
What's your four?	what kind of car are you driving?
What's your twenty?	what's your location?
Work twenty	location of employment
Zero hour	five minutes are up

The 10-code

The CB10 code was originally developed by American highway cops to avoid wasting time on the air. The code has largely been taken up by the CB users everywhere and so, to keep you fully abreast of the jargon of the motorways, here it is:

10-1	Poor reception
10-2	Good reception
10-3	Stop transmitting
10-4	OK message received
10-5	Relay message
10-6	Busy, stand by
10-7	Leaving air
10-8	In service, subject to call
10-9	Repeat message
10-11	Talking too rapidly
10-12	Visitors present
10-13	Advise weather and road conditions
10-16	Make pick-up at . . .
10-17	Urgent business
10-18	Anything for us?
10-19	Nothing for you, return to base
10-20	My location is . . .
10-21	Call by telephone
10-22	Report in person to . . .

10-23	Stand by
10-24	Completed last assignment
10-25	Can you contact . . .?
10-26	Disregard last message
10-27	I am moving to channel . . .
10-28	Identify your station
10-29	Time is up for contact
10-30	Does not conform to FCC rules
10-32	I will give you a radio check
10-33	Emergency traffic at this station
10-34	Trouble at this station, need help
10-35	Confidential information
10-36	Correct time is . . .
10-37	Pick-up truck needed at . . .
10-38	Ambulance needed at . . .
10-39	Your message delivered
10-41	Please tune to channel . . .
10-42	Traffic accident at . . .
10-43	Traffic jam at . . .
10-44	I have a message for you
10-45	All units within range please report
10-46	Help motorist
10-50	Break channel
10-60	What is next message number?
10-62	Unable to understand. Use landline
10-63	Network directed to . . .
10-64	Network clear
10-65	Awaiting your next message/ assignment
10-67	All units comply
10-69	Message received
10-70	Fire at . . .
10-71	Proceed with transmission in sequence
10-73	Speed trap at . . .
10-74	Negative
10-75	You are causing interference
10-77	Negative contact

10-81	Reserve hotel room for . . .
10-82	Reserve room for . . .
10-84	My telephone number is . . .
10-85	My address is . . .
10-89	Radio repairman needed at . . .
10-91	Talk closer to the mike
10-92	Your transmitter is out of adjustment
10-93	Check my frequency on this channel
10-99	Mission completed, all units secure
10-100	Stop at a lavatory
10-200	Police needed at . . .
10-2000	Dope pusher

ONE EON?

For a moment kindly discard your mantle of mathematical manipulative mastery and don your vestments of verbal virtuosity. I have a challenge for you.

Knowing that you can 'extract' a variety of numbers from a multiplicity of words – you can get TEN out of NET and EIGHT out of WEIGHT, for example – what I want you to do is work out what might be the *shortest* English words from which these numbers can be extracted. You can get ONE from WONDER, but can you get it from anything shorter?

one
two
three
four
five
six
seven
eight
nine
ten
eleven

twelve
thirteen
fourteen
fifteen
sixteen
seventeen
eighteen
nineteen
twenty
thirty
thirty-one
thirty-six
thirty-seven
thirty-nine
forty
forty-one
forty-nine
fifty
fifty-one
fifty-nine
sixty
sixty-three
seventy
eighty
eighty-one
eighty-nine
ninety
ninety-one
ninety-eight
ninety-nine
one hundred

(Words like 'fiver' and 'elevens' cannot be included since they are too similar to the numbers themselves. Answers, to the shortest I've found, are at the back.)

THE JOY OF SEX

There's more to SEX than sex.

The three letters SEX appear in plenty of other words that have nothing at all to do with sex. In many of these words, the SEX part is in some way indicative of the numbers 6, 16, 60, 600 or even 6000. There are also several words where the letters SEX appear and where neither sex nor numbers are involved.

Below you will find definitions for 25 words that all begin with 'sex'. What you have to do is to decide which of the *30* 'sex . . . ' words that follow apply to the *25* definitions below. Of course, five of them won't. (Answers at the back.):

The definitions
1 Six-year-old child
2 Stanza of six lines
3 Six-fold
4 To multiply by six
5 The office of a sexton
6 Piece of paper cut six from a sheet
7 Multiplied by sixty or a power of sixty
8 Having six fingers
9 Having six leaves
10 Relating to sixteen
11 Body of six colleagues
12 Divisions into sixths
13 Occurring every six years
14 Divided into six segments
15 Bronze coin of the Roman Republic
16 Having a chemical valency of six
17 Based on the number sixty
18 Danish dance for six couples
19 Person with six fingers
20 Of six thousand years
21 Measured by sixty degrees
22 A period of six years

23 Sectarian
24 Long fishing boat propelled by six oars
25 Proceeding by sixties

The words

sextoncy
sexfoil
sexfid
sexdigital
sexern
sexagene
sexennarian
sexious
sexavalent
sexto
sexagecuple
sextula
sextactic
sextuple
sexmillenary

sexadecimal
sextans
sextry
sextipartition
sexvirate
sextain
sextile
sexennium
sexennial
sexagenary
sexcuple
sextur
sexdigitist
sextant
sext

Afterword

Look at this sum:

$$11 + 2 - 1 = 12$$

It seems straightforward, doesn't it? But did you know that you could have obtained the same result by using the letters of the numbers instead?

By adding the letters of ELEVEN to the letters of TWO, and then by subtracting the letters of ONE, you would have ended up with the same result, the letters of TWELVE!

(ELEVEN + TWO = EEELNOTVW
EEELNOTVW − ONE = EELTVW = TWELVE)

And did you know that you can work out the number of books in both the Old Testament and the New

Testament by counting the number of letters in their names, and using the numbers you get? This is how you do it:

OLD = 3 + TESTAMENT = 9 = 39
NEW = 3 × TESTAMENT = 9 = 27

Divine, isn't it?

KEEP COUNTING

COUNT ON LOVE

The great American wit and versifier, Willard Espy, composed this delightful counting song to take you from one to ten and back again. To appreciate it fully, read it out loud.

1 Dear ewe, dear lamb, I've 1 thee: we
2 Will 2tle through the fields together;
3 With 3d and pipe we'll jubilee;
4 We'll gambol back and 4th in glee;
5 If 5 your heart, who gives a D.
6 How raw and 6 the weather?
7 In 7th heaven me and thee
8 Will 8 and soon find ecstasy –
9 My ewe be9 is tied to me;
10 And 10der is the tether!
10 Yet there may be a 10dency
9 (Someday when 9 no more know whether
8 You 8 for me still longlingly
7 Or find our love less 7ly)
6 For you in class 6 sulk to flee –
5 Off 5 no doubt to dew bellwether
4 And fresher 4 age. . .It may be
3 We'll both 3quire a style more free,
2 And find the Bird of Love 2 be
1 Reduced to 1 Pinfeather.

LETTERS COUNT

Here are four simple sums in which the letters stand for numbers. Can you work out what the numbers are? (Answers at the back.)

```
  ZQYOO           TOT
  PZJOO          +JOT
 +JYOOO           OJJ
  OZDQZ
                 WNYW
 AABHAEE         WWLL
 AAAWMEE         WYYW
+AAMHAEE        +WLSL
 XIURWMA         YYIAY
```

COUNT UP

a) Can you construct two addition sums (one of five digits, the other of four digits) using the nine digits from one to nine once each, so that both sums produce the same total?

b) Study this sum carefully and work out why the answer is wrong and why Lewis Carroll could give you a clue to the answer.

```
    3 И1 И
    3 И0
    7 И8 1 3
 И3 3 7 И8 1 3
```

c) With five odd-numbered digits, try to write an addition sum that will give you a total of 14.

d) Using the digits from one to seven write an addition

85

sum that comes to 100. In the calculation each of the digits may be used only once.

e) Try writing ten, first with five nines in an addition sum, and then with five nines in a subtraction sum.

f) Now see if you can write 100 using just four nines.

(Answers at the back.)

LAWS AND NUMBERS

LAYING DOWN THE LAWS

I have been collecting Laws and Axioms and Rules ever since I came across my first, which is still my favourite.

Cole's Law
Sliced cabbage.

I have others I like almost as much:
Bicycling, First Law of
No matter which way you ride, it's uphill and against the wind.

Billing's Law
Live within your income, even if you have to borrow to do so.

Local Anaesthesia, Law of
Never say 'oops!' in the operating theatre.

Newton's Unknown Law
A bird in the hand is safer than two overhead.

Pipe, Axiom of
A pipe gives a wise man time to think and a fool something to stick in his mouth.

Thurber's Conclusion
There is no safety in numbers, or in anything else.

Marginally more sophisticated, and definitely more relevant, are the next 20, each one of which should ring a chord – or strike a bell – with the number-conscious.

Ashley-Perry Statistical Axioms
1 Numbers are tools, not rules.
2 Numbers are symbols for things; the numbers and the things are not the same.
3 Skill in manipulating numbers is a talent, not evidence of divine guidance.
4 Like other occult techniques of divination, the statistical method has a private jargon deliberately contrived to obscure its methods from non-practitioners.
5 The product of an arithmetical computation is the answer to an equation; it is not the solution to a problem.
6 Arithmetical proofs of theorems that do not have arithmetical bases prove nothing.

Computer Programming Principles
1 The computer is never wrong.
2 The programmer is always wrong.

Coomb's Law
If you can't measure it, I'm not interested.

Finagle, The Law of the Too, Too Solid Point
In any collection of data, the figure that is most obviously correct – beyond all need of checking – is the mistake.

Finagle's Contributions to the Field of Measurement
1 Dimensions will be expressed in least convenient terms, e.g. Furlongs per (Fortnight)^2 = Measure of Acceleration.
2 Jiffy – the time it takes for light to go one cm in a

vacuum.

3 Protozoa are small, and bacteria are small, but viruses are smaller than the both of 'em put together.

Horowitz's Rule
A computer can make as many mistakes in two seconds as 20 men working 20 years.

Horowitz's Song for In-House Computer Programmes
'I/O, I/O, it's off to work we go. . .'

Jones's Mathematical Law
Twice nothing is still nothing.

Loderstedt's Rule
Measure twice because you can only cut once.

Mark's Law of Monetary Equalisation
A fool and your money are soon partners.

Metric Conversion, problem of
If God meant us to go metric, why did he give Jesus 12 disciples?

Murphy's Laws of Analysis
1 In any collection of data, the figures that are obviously correct will contain errors.
2 It is customary for a decimal to be misplaced.
3 An error that can creep into a calculation, will. Also, it will always be in the direction that will cause the most damage to the calculation.

Murphy's Laws, another
If mathematically you end up with the incorrect answer, try multiplying by the page number.

Pearson's Principle of Organizational Complexity
The difficulty in running an organization is equal to the square of the number of people divided by the sum of their true applied mentalities.
E.G. Normal Individual:

$$\frac{1^2}{1} = {}^1$$

Family of four (one teenager, one child):

$$\frac{4^2}{1 + 1 + .5 + .3} = 5.71$$

$$\frac{\text{Many}^2}{13.2} = \text{Infinity}$$

Penner's Principle
When the maths start to get messy – Quit!

Shelton's Laws of Pocket Calculators
1 Rechargeable batteries die at the most critical time of the most complex problem.
2 When a rechargeable battery starts to die in the middle of a complex calculation, and the user attempts to connect house current, the calculator will clear itself.
3 The final answer will exceed the magnitude or precision or both of the calculator.
4 There are not enough storage registers to solve the problem.
5 The user will forget mathematics in proportion to the complexity of the calculator.
6 Thermal paper will run out before the calculation is complete.

Spencer's Laws of Accountancy
1 Trial balances don't.
2 Working capital doesn't.
3 Liquidity tends to run out.
4 Return on investments won't.

Twyman's Law
Any statistic that appears interesting is almost certainly a mistake.

Wain's Conclusion

The only people making money these days are the ones who sell computer paper.

Wallace's Two-out-of-three Theory

Speed
Quality
Price
Pick any two

MYSTERIES OF MULTIPLICATION

AMAZING MULTIPLIERS

Here's a handy way of multiplying by five, 25 and 125. As you know, placing a decimal point in front of each of these digits turns them into decimal fractions, which can be expressed as ordinary fractions:

    .5 = 1/2
    .25 = 1/4
    .125 = 1/8

This means that you could write five, 25, and 125 as:

    5 = 10/2
    25 = 100/4
    125 = 100/8

The purpose of showing you this is to demonstrate that multiplying by any of these numbers is really the same as multiplying by either ten, 100, or 1000, and then dividing by either two, four or eight. So instead of performing a lengthy multiplication, in the case of 125, for example, all you have to do is to add three zero's and then divide by eight. Watch:

a) 754696×5 $\dfrac{3773480}{2/7546960}$
   $= 7546960 \div 2$

b) 754696×25 $\underline{18867400}$
          $= 75469600 \div 4$ $= 4/75469600$

c) 754696×125 $\underline{94337000}$
          $= 75696000 \div 8$ $= 8/754696000$

And here's an equally ingenious method for multiplying the teens, the numbers between ten and 20:
1 Add the first number to the *second* digit of the second number.
2 Now multiply that by ten.
3 Multiply the second digits of both numbers.
4 Finally add the two answers and you'll get the answer to the multiplication sum.
Here goes:

$17 \times 19 = ?$
1 $\quad 17 + 9 = 26$
2 $\quad 26 \times 10 = 260$
3 $\quad 7 \times 9 = 63$
4 Add 260 $\underline{260}$
          323

$15 \times 11 = ?$
1 $\quad 15 + 1 = 16$
2 $\quad 16 \times 10 = 160$
3 $\quad 5 \times 1 = 5$
4 Add 160 $\underline{160}$
          165

MULTIPLICATSKY

Before more conventional, Western methods were introduced into the Soviet Union, Russian peasants used a novel system of multiplication that only required the ability to double a number, halve it and perform simple addition. The sum itself took the form of two columns of figures, headed by the ones they wished to multiply. The calculation then proceeded like this:

1 Step by step the numbers in the left-hand column would be halved while the ones in the right-hand column were doubled. This process ignored any remainders in the left-hand column and carried on until one was reached.

2 All the even numbers in the left-hand column would now be crossed out, together with their corresponding numbers in the right-hand column.

3 The final stage simply involved adding the remaining numbers in the right-hand column to obtain the answer to the multiplication sum.

Here's an example 87 × 56:

1 Form columns and halve/double

$$87 \times 56$$
$$43 \times 112$$
$$21 \times 224$$
$$10 \times 448$$
$$5 \times 896$$
$$2 \times 1792$$
$$1 \times 3584$$

2 Cross out all even numbers and adjacent ones

$$87 \times 56$$
$$43 \times 112$$
$$21 \times 224$$
$$\cancel{10} \times \cancel{448}$$
$$5 \times 896$$
$$\cancel{2} \times \cancel{1792}$$
$$1 \times 3584$$

3 Add remaining numbers in right-hand column. Answer

$$= \quad 4872$$

NEVER ODD
OR EVEN

PALINDROMES

A verbal palindrome is a word or a phrase or a sentence that reads the same forwards as backwards – like 'nun' or 'peep' or 'repaper' or 'Never odd or even.'

In the world of numbers palindromes appear too. There are straightforward ones – 37873 for example – and much more remarkable ones that appear in certain calculations, like these:

$$10989 \times 9 = 98901$$
$$21978 \times 4 = 87912$$

More sophisticated numerical palindromes involve two calculations, like these:

$9 + 9 = 18$	$81 = 9 \times 9$
$24 + 3 = 27$	$72 = 3 \times 24$
$47 + 2 = 49$	$94 = 2 \times 47$
$497 + 2 = 499$	$994 = 2 \times 497$

For an even more complex way of having fun with a number palindrome try adding its reverse to any number, over and over again, until you finally achieve a palindrome in the result. This is what I mean:

```
      38
  +  83
    121
```

```
      139
  +  931
     1070
  +  0701
     1771
```

```
   48,017
  + 71,084
   119,101
  + 101,911
   221,012
  + 210,122
   431,134
```

Be warned: some numbers are easier to play with than others. For example, computers have failed to find a palindrome for 196 even after performing this step thousands and thousands of times!

And while we're on the subject, here's a timely question. Can you think of a palindromic way of expressing the number 12?

LUCKY NUMBERS

The oldest verbal palindrome must be the one first uttered in the Garden of Eden shortly after the Creation: 'Madam, I'm Adam.' The most famous must be the Emperor Napoleon Bonaparte's lament, 'Able was I ere I saw Elba.'

Napoleon's lucky number, by the way, was four. What's yours? If you don't know, your name will tell you. Each letter of the alphabet has a value:

A = 1	H = 8	O = 15	V = 22
B = 2	I = 9	P = 16	W = 23
C = 3	J = 10	Q = 17	X = 24
D = 4	K = 11	R = 18	Y = 25
E = 5	L = 12	S = 19	Z = 26
F = 6	M = 13	T = 20	
G = 7	N = 14	U = 21	

To work out your own number, write out your name and put the value of the letters underneath:

W I L L I A M
23 9 12 12 9 1 13

S H A K E S P E A R E
19 8 1 11 5 19 16 5 1 18 5

Now add up all the figures:

23 + 9 + 12 + 12 + 9 + 1 + 13 + 19 + 8 + 1 + 11 + 5 + 19 + 16 + 5 + 1 + 18 + 5 = 187

Now take the digits in the answer to the sum and add *them* together:

1 + 8 + 7 = 16

Add the digits from that answer together:

1 + 6 = 7

And you have the number you need. Seven, it turns out, is William Shakespeare's lucky number.

According to followers of the pseudo-science of numerology – which is as old as Pythagoras who, in the sixth century BC, laid great emphasis on the numerical harmony of the universe – your lucky number is of enormous significance. It leads you to your lucky day of the week:

1 Saturday, Wednesday
2 Sunday

3 Thursday
4 Monday
5 Tuesday
6 Monday, Friday
7 Thursday
8 Friday
9 Monday

And to your lucky colour:

1 Black
2 Yellow
3 White
4 Blue
5 Green
6 Grey
7 Purple
8 Brown
9 Red

Most revealing of all, your lucky number indicates your dominant characteristics:

1 Self-concern, enterprise, ambition
2 Passivity, receptivity
3 Versatility, optimism
4 Practicality, unimaginativeness, endurance
5 Victory, instability, sexuality
6 Domesticity, dependability
7 Knowledge, solitariness
8 Teamwork, organization, material success
9 Health, achievement

I don't believe a word of it, of course, and nor do you. Yet we still want to know our own lucky numbers. Odd, isn't it?

ODDS ON

ROULETTE – THE ODDS AGAINST YOU

In 1891 Charles Deville Wells took himself off to Monte Carlo with £400 in his pocket. Three days after entering the casino there he emerged with £40,000 (worth well over £800,000 today). Wells was the original 'man who broke the bank at Monte Carlo'. We will not see his like again, for two very good reasons: the introduction of the zero on the roulette wheel and the house limit on stakes at any single spin.

The existence of a zero loads the odds in favour of the casino. If the ball lands on zero, all the bets on the table are wiped out, and in consequence the odds swing in favour of the casino to the tune of 1 1/18 to 1.

When Charles Wells made history in 1891, there was no zero and he worked on the principle of doubling his stake every time he lost. The process of doubling up, as it's called, is a long-winded way of winning – and it's risky. If you started out with a £1 stake doubled after every loss, and lost 30 spins in a row, you would then owe the bank a cool £1,073,741,823!

If there is any way to beat the house at roulette, it's probably in paying careful attention to the mechanics of the wheel itself. At the end of the last century another intrepid Englishman, William Jaggers, decided to test his theory that no roulette wheel could be perfectly balanced so that careful observation would show up its bias. He took himself off to Monte Carlo for a little field work, found a table that suited him and was banned from the establishment for life when he won £80,000!

HELPING HANDS

In a normal 52-card game of poker, there are nine commonly accepted combinations of cards that score. Here they are with a tenth option, bust, plus the number of these hands, the probability of their appearance and the odds against them showing up:

Type of hand	Number of these hands	Probability (%)	Approximate odds against drawing
Royal flush	4	0.00015	649,739 to 1
Straight flush	36	0.0014	72,192 to 1
Four of a kind	624	0.024	4,164 to 1
Full house	3,774	0.14	693 to 1
Flush	5,108	0.20	508 to 1
Straight	10,200	0.39	254 to 1
Three of a kind	54,912	2.11	46 to 1
Two pair	123,552	4.75	20 to 1
One pair	1,098,240	42.26	4 to 3
Bust	1,302,540	50.12	even

And here's the same prediction for a 13-card bridge hand:

Type of hand	Number of these hands	Probability (%)	Approximate odds against drawing
Four aces	1,677,106,640	0.26	378 to 1
Two aces	192,928,249,296	30.4	7 to 3

Type of hand	Number of these hands	Probability (%)	Approximate odds against drawing
No card higher than a nine (Yarborough)	347,373,600	0.055	1,827 to 1
Seven-card suit or longer	25,604,567,408	4.03	24 to 1
Any 4-3-3-3 distribution	6.691×10^{10}	10.54	17 to 2
Any 5-4-2-2 distribution	6.718×10^{10}	10.58	17 to 2
Any 5-3-3-2 distribution	9.853×10^{10}	15.52	11 to 2

DICE

The earliest known cubical dice were found in tombs in Egypt and date from before the second millenium BC. Playing with dice was very popular in the ancient world and by the Middle Ages it had become so popular throughout Europe that there were dicing schools and guilds established to cater for the game. If you want to know the various ways in which dice can fall and the probabilities of those falls actually happening, the table on page 102 will show you:

BIRTHDAY COINCIDENCES

Go into a room where there are 30 people chosen at random and there is a 70 per cent chance that there will be two people in the room who share the same birthday, probably without knowing it. As the number of people in the room rises, so does the likelihood of two

Sum	Ways to make		Chances
2	1	⚀⚀	1/36
3	2	⚀⚁ ⚁⚀	1/18
4	3	⚀⚂ ⚂⚀ ⚁⚁	1/12
5	4	⚀⚃ ⚃⚀ ⚁⚂ ⚂⚁	1/9
6	5	⚀⚄ ⚄⚀ ⚁⚃ ⚃⚁ ⚂⚂	5/36
7	6	⚀⚅ ⚅⚀ ⚁⚄ ⚄⚁ ⚂⚃ ⚃⚂	1/6
8	5	⚁⚅ ⚅⚁ ⚂⚄ ⚄⚂ ⚃⚃	5/36
9	4	⚂⚅ ⚅⚂ ⚃⚄ ⚄⚃	1/9
10	3	⚃⚅ ⚅⚃ ⚄⚄	1/12
11	2	⚄⚅ ⚅⚄	1/18
12	1	⚅⚅	1/36

102

of them sharing a birthday. But as the number falls, to say 23 people, the probability of two of them having the same birthday drops to 0.5073. This graph charts the increasing probability of the shared birthday from the slim chance, when there are no more than ten people present, to the near certainty when there are six times that number collected together.

Probability

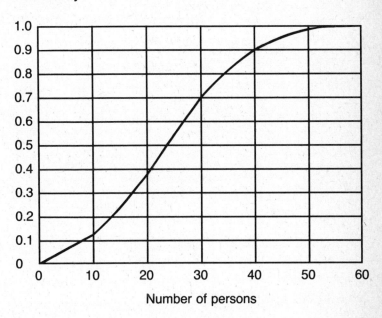

Number of persons

PUZZLING NUMBERS

THE MISSING NUMBERS

There's a number missing in each of these puzzles.
Can you work out what each one should be?

a)

9	4	6	60
11	2		32
80	1	2	82

d)

632	36	981	72
333		463	72
704	0	212	4

b)

6	0	4
3		2
1	5	4

e)

16	6	1	30
12	9	11	9
3	6	22	
4	4	2	10

c)

63	54	45	36
94	81	72	63
79		54	45

f)

94	11	83
62		13
76	53	23

CHALLENGING NUMBERS

1 My first wife – I'm still married to her, but I call her that to keep her on her toes – entered a department store with a fixed sum of money and a definite plan of how she was going to spend it, which was as follows:

$\frac{1}{10}$ on shoes
$\frac{4}{9}$ of the remainder on clothes
$\frac{1}{10}$ of the remainder on a scarf
$\frac{2}{9}$ of the remainder on a hat
$\frac{4}{7}$ of the remainder on stationery
$\frac{2}{3}$ of the remainder on chocolates
$\frac{49}{50}$ of the remainder on a bracelet

On leaving the store, she gave five tenths of the remainder to a busker and on the way home she bought herself a cup of tea with four fifths of the money she had left. When she got back to Brandreth Towers she had just one penny left. How much did she set out with?

2 I am told the Charles is twice as old as Martin used to be when Charles was as old as Martin is now. Martin is 18. How old is Charles?

3 I've found that this number reduced by six and the remainder multiplied by six, will give the same result, if you reduced this number by nine and multiplied the remainder by nine. What's the number?

4 Take a seventh of the result you get by adding 12 to this number, and you'll discover that it is twice the result you get from subtracting one from the number. What is the number?

5 Travelling between A and B there are 24 miles uphill, 36 miles downhill and 12 miles along the flat. A courier takes 25 hours to go and 28 hours to return, but if, while going from A to B, there had been 12 miles uphill, 24 miles level (and 36 miles downhill), he could travel the distance in 23 hours. What is his rate per hour uphill, downhill and on the level?

6 The difference between the squares of two numbers is 16, and one number is exactly three fifths of the other.
What are the numbers?

7 If a quarter of 20 is not five but four, then what should a third of ten be?

8 I was offered the choice of one of each type of rosebush out of a collection of 12. There were five red roses and seven white. How many possible choices did I have to select one of each?

9 A rope stretches from the top of one pole 50 feet high to the top of another pole 20 feet high, standing 16 feet away. How long is the rope?

10 The fish's tail weighs nine pounds. Its head weighs as much as the tail and a third of the body combined, and the body weighs as much as the tail and the head combined. How much does the fish weigh?

11 Two classrooms can hold 105 pupils. There are 60 boys and 45 girls. If ten per cent of the boys and 33⅓ per cent of the girls usually stay absent over a period of time, what would be the percentage of absenteeism for the whole?

12 Bill can run a mile in 4·12 minutes, but Ben can run 4·12 miles in an hour. Which is the faster of the two?

13 Add eight to this number, subtract eight from the sum, multiply what's left by eight, divide the product by eight and you're left with four. What's the number you started with?

14 Whenever this man sits down to write a letter he always uses two sheets of paper and for every 12 letters that he writes, he throws away one sheet. Supposing that he has 100 sheets to use, how many letters could he write with them?

A QUESTION OF AGE

'If half of half my age,' said Kate,
'Were now replaced by twenty-eight,
'And then the whole reduced by four,
'You'd make my age just eighteen more.
'Why, fancy! When I'm forty-six,
'My age will be just double Dick's,'
That's what she said, and he's her son:
So figure out his age for fun.

CHINESE CHALLENGE

Here is a puzzle that is some 5000 years old!

If:

= 6

= 1

= 3

What does: represent?

QED

THINK OF A NUMBER

Latin was not my best subject at school – for years I thought *In loco parentis* meant *My dad's an engine driver* – but I do know that *Quod erat demonstrandum* means roughly *Which was the thing to have been proved.* I know it because it was the line the maths masters at school would declare triumphantly at the conclusion of one of the many tricks that were his speciality. He knew a dozen ways of telling you 'the number you first thought of' – and here are three of them.

In the first trick you ask your victim to:

a) Think of a number
b) Double it
c) Add four
d) Multiply by five
e) Add 12
f) Multiply by ten
g) Tell you the answer

Now in your head you subtract 320, cross off the last two digits, and, hey presto, you've got the number your victim thought of. Announce it with pride – adding 'QED' if you're so inclined.

Here's how the trick works:

a) Think of a number, say 54   =   54
b) Double it   =   108
c) Add four   =   112
d) Multiply by five   =   560
e) Add 12   =   572
f) Multiply by ten   = 5720
g) Get the answer, subtract 320 and
cross off the last two digits   =   54

Here's the second trick:
a) Get your victim to think of a number
b) Ask him or her to square it
c) Ask him or her to subtract one from the number he
or she first thought of
d) Ask him or her to square this second number
e) Ask for the difference between the two squares
f) Now in your head add one to this number
g) Still in your head halve the answer and you'll have
the number your victim thought of

Here's how it works:

a) Think of a number, say 26   =   26
b) Square it   = 676
c) Subtract one from 26   =   25
d) Square 25   = 625
e) Ask for the difference between the two
squares   =   51
f) Add one   =   52
g) Halve the answer and tell your
victim the number he or she first
thought of   = 26!

QED

And here's the third trick. It's an easy one to perform
because the answer's always seven.

a) Ask your victim to think of a number
b) Ask him or her to double it

c) Ask him or her to add five
d) Ask him or her to add 12
e) Ask him or her to subtract three
f) Ask him or her to divide the total by two
g) Ask him or her to subtract the number he or she first thought of
h) Tell your victim that the answer is seven

And this is how it works:

a) Think of a number, say 84	=	84
b) Double it	=	168
c) Add five	=	173
d) Add 12	=	185
e) Subtract three	=	182
f) Divide by two	=	91
g) Subtract the number first thought of (84)	=	7
h) Announce that the number is seven		

QED

BOX CLEVER

This is a rather different kind of trick and appears to involve an astonishing feat of memory. In fact, all it requires is the ability to remember a set of simple instructions and perform a little elementary addition.
This is what happens:
Show your audience the grid on page 112. They will see that it contains 36 boxes (numbered from one to 36) and that each box contains a different nine-digit number. Explain to your audience that yours is an amazing memory and that you know which nine-digit number is in which box. Invite one member of the audience to hold the grid so that you cannot possibly see it. Then ask them to pick a box and give you the box number.

A moment after you've been given the box number,

you can announce the nine-digit number in that box and take your well-earned applause.

This is how you do it:
1 In your head, add 11 to the box number you've been given
2 Reverse the digits of this sum to get the first two digits of the nine-digit number
3 Add these digits to get the third digit. Drop any tens
4 Add the last two digits in your total and drop any tens. This will give you your fourth digit of the nine-digit number
5 Continue to add the last two digits in your total, discarding any tens, until you arrive at a nine-digit number
6 Announce the nine-digit number with pride

Here's an example of how the system works:

1 Say Box 35 is chosen.
2 Add 11 to 35 = 46
3 Reverse these digits to get the first two of the large number in box 35   = 64
4 Add these to get the third digit of the large number: $6 + 4 = 10$ 640
5 Add the last two digits above to get the fourth digit: $4 + 0 = 4$ 6404
6 Add the last two digits etc. $0 + 4 = 4$ 64044
7 Add the last two digits etc. $4 + 4 = 8$ 640448
8 Add the last two digits etc.
$4 + 8 = 12$ (drop tens 12)   6404482
9 Add the last two digits etc.
$8 + 2 = 10$ (drop tens 10)   64044820
10 Add the last two digits etc. $2 + 0 = 2$ 640448202
11 You've got nine digits – add the number in box 35

QED

23 437077415	39 055055055	18 921347189	22 336954932	4 516730336	38 943707741	16 729101123
2 314594370	45 651673033	30 145943707	34 549325729	25 639213471	6 718976392	15 628088640
9 022460662	37 842684268	46 752796516	3 415617853	1 213471897	17 820224606	32 347189763
21 235831459	5 617853819	44 550550550	11 224606628	41 257291011	19 033695493	8 910112358
29 044820224	12 325729101	33 448202246	13 426842684	43 459437077	7 819099875	10 123583145
49 066280886	14 527965167	24 538190998	47 853819099	26 730336954	40 156178538	28 932572910
31 246066280	27 831459437	35 640448202	48 954932572	20 134718976	42 358314594	36 741561785

THE THREE DIGITS

Here is a trick that's (fairly) pure and (relatively) simple:

a) Ask a member of your audience to think of a three digit number in which all the numbers are the same.

b) While your victim is doing this, jot down a number on a piece of paper, fold it and pass it to another member of the audience.

c) Now ask your victim to add the digits in the number he or she has thought of.

d) Tell your victim to divide the orginal three-digit number by the sum of the digits.

e) Ask your victim for the answer.

f) Now ask the member of the audience holding your piece of paper to look at it and read out the number on it. It will be 37.

Here's how the trick works:

a) Three digit number, say 555   555

b) Jot down 37 on piece of paper and fold to hide the number

c) Add the digits ($5 + 5 + 5 = 15$)   15

d) Divide 555 by 15 ($555 \div 15 = 37$)   37

e) What does the paper say?   37

QED

THE MISSING DIGITS

In this trick you'll be able to impress your victim with your uncanny ability to name a particular digit dropped arbitrarily from a list of digits chosen at random and then rearranged.
This is what you do:

a) Ask your subject to write down *any number* he or she chooses.

b) Ask him or her to add all the digits in that number.

c) Ask him or her to subtract the sum of the digits from the original number.

d) Ask him or her to jumble the digits and rewrite them in any order.

e) Ask him or her to add 25 to this new number.

f) Ask him or her to cross out one digit, but not a zero.

g) Ask for the number that is left.

Once you've written this down you:

a) Add together the digits.

b) Subtract seven from this total.

c) Subtract this number from the next multiple of nine.

d) Announce that the answer you have left, is the digit that was crossed out.

And here is an example of the way it works:

a) Write down any number at random, say = 7529874

b) Add all the digits (7+5+2+9+8+7+4)   = 42

c) Subtract this from 7529874   = 7529832

d) Jumble and rewrite digits, say   = 9872253

e) Add 25   = 9872278

f) Cross out one digit, but not a zero   9872278

g) Ask for number   = 972278

h) Add digits (9+7+2+2+7+8)   = 35

i) Subtract seven from this total   = 28

j) Subtract 28 from next highest multiple of nine (36)   = 8

k) Announce that the number crossed out was   8

QED

Finally, here's another version of the same trick, but it's one you should be able to master without recourse to pencil and paper. This is what you do:

a) Ask your victim to think of a number.

b) Ask your victim to multiply this by 100.

c) Ask your victim to add 36.

d) Ask your victim to subtract the number first thought of.

e) Ask for the digits in the total *except for one of them*.

This is what you then do:

a) Add all these digits.

b) Subtract their sum from the next highest multiple of nine.

c) Announce the missing digit, which is the answer to your subtraction.

Here it is in practice:

a) Think of a number, say 73421   = 73421
b) Multiply by 100   = 7342100
c) Add 36   = 7342136
d) Subtract 73421   = 7268715
e) Obtain the digits except for one   = 7268/15
f) Add these digits (7+2+6+8+1+5)   = 29
g) Subtract 29 from next highest multiple
of nine (36)   = 7
h) Announce that the missing digit was indeed   7
QED

DICEY

This trick involves dice, but it's no gamble. Invite your victim to roll three dice and to make sure you don't see them. When the dice have been rolled, ask your victim to undertake some minor calculations, at the conclusion of which you will be able to tell him which three dice he rolled:

This is what you ask your victim to do:

a) Roll three dice, out of your sight.

b) Double the number of dots on the first die.

c) Add three.

d) Multiply that sum by five.

e) Add the number of the second die.

f) Multiply this sum by ten.

g) Add the number of dots on the third die.

h) Tell you the total.

Once you have the total all you have to do is:

a) Take the total and subtract 150, leaving a three-digit number.

b) The first digit will be the same as the number of dots on the first die.

c) The second digit will be the same as the number of dots on the second die.

d) And the third digit will be the same as the number of dots on the third die.

e) Tell your subject the sequence of numbers which was shown by the rolled dice.

If the rolled dice showed four, five and two this is how the trick would work:

Double the dots on the first die (four)	= 8
Add three	= 11
Multiply by five	= 55
Add number of dots on second die (five)	= 60
Multiply by ten	= 600
Add number of dots on third die (two)	= 602
Ask for total	= 602
Subtract 150	= 452

QED

NAME THE DAY

Tell your audience that you know the precise days of the week for every single day in the nineteenth and twentieth centuries! Invite someone to name a particular date and then tell them what day of the week it was or will be on that date.

To perform this trick you have got to remember key

numbers for each month of the year. To help you the distinguished American number wizard has described a series of ingenious mnemonics:

Month	Key number	Mnemonic
January	1	The FIRST month
February	4	A C-O-L-D month (four letters)
March	4	The K-I-T-E month
April	0	On April Fool's Day I fooled NObody
May	2	'May Day' is TWO words
June	5	The B-R-I-D-E month
July	0	On July 4 I shot NO firecrackers
August	3	The H-O-T month
September	6	Start of A-U-T-U-M-N
October	1	A witch rides ONE broom
November	4	A C-O-O-L month
December	6	Birth of C-H-R-I-S-T

Assuming that you've managed to learn the key numbers for each month, this is how you progress:

1 Take the last two digits of the year and consider them as a single number.
2 Add the number of dozens in that number.
3 Add the remainder.
4 Add the number of times four goes into the remainder, if in the 19th century, add two.
5 If the result is seven or greater than seven, divide by seven and remember only what is left.
6 Now add the key number of the given month. Again if this result is seven or greater, divide by seven and remember only what is left.
7 Now add the day of the month. As before, if the result is seven or greater, divide by seven and remember only what is left.

8 The digit you are left with will give you the day of the week if you count:

Saturday as 0	Wednesday as 4
Sunday as 1	Thursday as 5
Monday as 2	Friday as 6
Tuesday as 3	

9 If the year happens to be a Leap year, and the month is either January or February, go back one day for your final result.

And how do you recognise a Leap year? Easy – the first stage will indicate this. Leap years are multiples of four, and any number can be divided by four if its last two digits can. So, if there is no remainder when you divide by 12, or none when you divide by four, you know that you're dealing with a Leap year. (And the exceptions? 1800 and 1900. These can be divided by four, of course, but they're *not* Leap years. The year 2000 is a Leap year, however.)

So let's imagine that you have been asked to name the day on which the first manned landing took place on the Moon.

20th July 1969. This is how it's done:

1 Take the last two digits 1969

2 Divide by 12 = 5 remainder 9

3 Divide remainder by 4 = 2 (remainder 1)

4 Add three parts 5 + 9 + 2 = 16

5 16 is greater than 7 so divide by 7 = 2 remainder 2

6 Add key number for July (0) 2 + 0 = 2

7 Add day of month 2 + 20 = 22

8 22 is greater than 7 so divide by 7 = 3 remainder 1

9 1 corresponds with SUNDAY on the week chart

10 So 20th July 1969 was a Sunday.

QED

HAPPY BIRTHDAY TO YOU

With the next five tricks you must choose your victims with care. Not everyone likes to have their age revealed.

With the first trick, you are going to ask your victim to perform a series of calculations in his or her head.

a) First get your victim to think of a number between one and nine.

b) Ask him or her to double it.

c) Ask him or her to add five.

d) Ask him or her to multiply the result by 50.

e) Ask him or her to add the number of the current year, less 250 (in 1984, therefore, add 1984 − 250 = 1734).

f) Ask him or her to subtract the year of his or her birth.

g) Ask him or her to tell you the number he or she has arrived at. It will be a three-digit number. The first digit will be the number the subject first thought of and the second two digits will be his or her present age.

Here's an example:

a) Your victim chooses eight.

b) Double this   = 16

c) Add five   = 21

d) Multiply by 50   = 1050

e) Add the year you are in at present less 250 (1984 − 250 = 1734)   = 2784

f) Subtract the year of the victim's birth, say 1934   = 850

g) Study the answer. The first digit is the one first thought of, the last two are his or her age this year.

QED

With this next trick you can reveal to your victim the

day of the week, the month in the year and the date in that month on which he or she was born – as well as how old he or she is. This is what you do:

a) Ask your victim to write down a number which shows the day of the week, the month in the year and the date in that month in which he or she was born. The days are counted from one to seven starting with Monday, and the months, of course, are counted from one to 12, starting with January.

b) Get your victim to double the number.

c) Add five.

d) Multiply by 50.

e) Add his or her present age.

f) Subtract 365.

g) Add 115 and tell you the answer.

The answer will give you:

a) The day of the week on which your victim was born, shown by the first digit.

b) The month in which the subject was born, shown by the second and third digits.

c) The date in the month on which he or she was born, shown by the next digit, or digits.

d) And finally your subject's present age, shown by the last digits, or digit.

To see how it works in practice, let's imagine that our victim was born on 6th September 1919, which was a Saturday:

a) The number written down will be 696 – six for Saturday, nine for September and six for the date in September of the birthday.

b) Double this   = 1392
c) Add five   = 1397
d) Multiply by 50   = 69850
e) Add the subject's present age (65 in 1984) = 69915
f) Subtract 365   = 69550

g) Add 115 and look at your answer   = 69665

Reading from left to right, you got the day of the week on which your subject was born (six = Saturday); the month in which he or she was born (nine = September); the date of the birthday (six); and his or her age this year (64).

QED

Trick No 3 in the birthday series is a simple one:
a) Ask your victim to jot down any random five-digit number.
b) Ask him or her to multiply it by two.
c) Add five.
d) Multiply by 50.
e) Add his or her age.
f) Add 365.
g) Subtract 615 and tell you the answer.

The answer will be a seven-digit figure of which the first five digits will be the original random number and the last two will be the subject's age.
See how it works in practice:

a) Pick a five-digit number, say 62981   = 62981
b) Multiply this by two   = 125692
c) Add five   = 125697
d) Multiply by 50   =6298350
e) Add the subject's age, say 36 =6298386
f) Add 365   =6298751
g) Subtract 615 and look at the answer =6298136

The first five digits are the same as the random number written at the top, while the last two reveal your victim's age.
QED

Trick No 4 again requires your victim to do all the work:

a) Ask your victim to write down his or her age.

b) Ask him or her to multiply this by three.

c) Ask him or her to add one.

d) Ask him or her to multiply this sum by three.

e) Ask him or her to add his or her age to this sum.

f) Ask him or her to subtract three.

g) Ask your victim to tell you the final answer, knock off the last digit, and you'll be left with his or her's age.

Here's an example:

a)	The victim writes down his or her age, say 57	= 57
b)	Multiply by three	= 171
c)	Add one	= 172
d)	Multiply this by three	= 516
e)	Add age, 57	= 573
f)	Subtract three	= 570
g)	Ask for total and knock off final digit	= 57$\emptyset$

QED

Finally here's a chart that will lead you to the age of anyone – so long as they are not more than 63 years old. (If they are, you shouldn't be so tactless as to be trying tricks like this out on them anyway.)

All you have to do is ask your victim to indicate which of the six columns contain their age. Then, in your head, you simply add the numbers at the top of the columns indicated, and the sum will give you your victim's age.

Here's an example:

a) Your victim is 57.

b) He or she indicates that his or her age appears in the columns headed by 32, 16, 8 and 1, columns a, b, c and f.

c) Add the numbers at the top of these four columns and you get a total of 57.

QED

(a)	(b)	(c)	(d)	(e)	(f)
32	16	8	4	2	1
33	17	9	5	3	3
34	18	10	6	6	5
35	19	11	7	7	7
36	20	12	12	10	9
37	21	13	13	11	11
38	22	14	14	14	13
39	23	15	15	15	15
40	24	24	20	18	17
41	25	25	21	19	19
42	26	26	22	22	21
43	27	27	23	23	23
44	28	28	28	26	25
45	29	29	29	27	27
46	30	30	30	30	29
47	31	31	31	31	31
48	48	40	36	34	33
49	49	41	37	35	35
50	50	42	38	38	37
51	51	43	39	39	39
52	52	44	44	42	41
53	53	45	45	43	43
54	54	46	46	46	45
55	55	47	47	47	47
56	56	56	52	50	49
57	57	57	53	51	51
58	58	58	54	54	53
59	59	59	55	55	55
60	60	60	60	58	57
61	61	61	61	59	59
62	62	62	62	62	61
63	63	63	63	63	63

RING ROUND THE WORLD

NUMBER OBTAINABLE

The numbers that loom largest in my life are telephone numbers. I collect them. Some of them I use regularly, many of them I never use at all, most of them I try to memorise and all of them I cherish.

Here are a few of my favourites. Thanks to Subscriber Trunk Dialling, to take advantage of them all you need is a Dialling Code book (to give you the relevant codes for the various countries and cities concerned), a working telephone and a bank balance that will allow you to foot the bill.

Teatro alla Scale	Milan 8879
Central Intelligence Agency	Washington DC 351-1100
Speaking Clock	Madras 174
Raffles Hotel	Singapore 328041
Moenjodaro (ancient city, Indus Valley)	Moenjodaro 6
Baku (table reservations)	Moscow 99-80-94
Acropolis	Athens 3236-665
British Embassy (Mongolia)	Ulan Bator 51033/4
City Alcoholic Treatment Centre	Chicago 254-3680
Karachi Race Course	Karachi 515809
Sherpa Society	Kathmandu 12412

Eiffel Tower	Paris 705-44-13
Ambulance	Bangkok 13
Wayside Chapel Crisis Centre	Sydney 33-4151
Warner Brothers Inc	Burbank 845-6000
Artificial Limb Centre	Auckland 600 644
The White House	Washington DC 456-1414
Entebee Airport	Entebee 2516
Port Stanley Aerodrome	Port Stanley 219
Weather Forecast	Bombay 211654
Air and Sea Rescue	Bahamas 2-3877
Part-Time Child Care	New York 879-4343
Wells Fargo & Company	San Francisco 386-0123
Jack the Clippers	Dublin 90-94-97
Folies Berjer	Istanbul 449569
Station Taxi rank	Salzburg 72680
Oil Can Harry's	Vancouver 683-7306
Pinkerton's, Inc	New York 285-4860
Colt Fire-arms	Hartford 278-8550
HM Queen Elizabeth II	London 930-4832
Yvonne's Beauty Centre	Rasal-Khaimah 29540
Oman Family Bookshop	Ruwi 702858
Tokapi Palace	Istanbul 28-35-47
Lobster Pot	Nairobi 20491
Yves Saint-Laurent	Paris 265-74-59
Hiroshima Castle	Hiroshima 21-7512
Anti-Imperialist Hospital	Peking 553-731
Kaziranga Wildlife Sanctuary, Assam	Kaziranga 3
Highgate Cemetery	London 340-1834
Wrigley de Mexico	Naucalpan de Juarex 5-76-58-53
Sunset Club	Lima 22-98-74
Paranormal, Investigation of claims of	Buffalo 837-0306
Dubai Broadcasting	Dubai 470255
Champagne J Bollinger SA	Reims 50-12-34
United States Supreme Court	Washington DC 252-300
Ministry of Education	Zomba 6111

Watergate	Washington DC 965-230
Dial-a-Girl	Sydney 357-1125
Greek Powder & Cartridge	
Company, Inc	Athens 3-236-091
Waneroo Football Club	Perth 409-9548
Nippon Steel Corporation	Tokyo 242-4111
Gstaad Golf Course	Gstaad 42636
Wig Corporation	Kaohsiung 220740/1
Dial-a-Joke	New York 999-3838
Horoscopes: Aries	New York 936-5050
Taurus	" " " -5151
Gemini	" " " -5252
Cancer	" " " -5353
Leo	" " " -5454
Virgo	" " " -5656
Libra	" " " -5757
Scorpio	" " " -5858
Sagittarius	" " " -5959
Capricorn	" " " -6060
Aquarius	" " " -6161
Pisces	" " " -6262
Information	Beirut 340940
The Raja of Perlis	Perlis 7552211
Kabul Airport	Kabul 25541-46
European Space Agency	Paris 567-5578
Organization of African Unity	Addis Abbaba 47-480
World Future Society	Bethseda 656-8274
Cartier Inc	New York 753-0111
Bulawayo Public Library	Bulawayo 6-09-66
Pleasure! Magazine	Accra 66640
Stanton the Magician	Perth 446-8772
Bondi Junction Health Centre	Sydney 389-5072
Jordan Phosphate Mines	
Company Ltd	Amman 38147
Big Mountain	Whitefish 862-3511
Scary Pictures Corporation	Toronto 925-0425
Delphi	Delphi 82313

Dial-an-Angel	Sydney 467--5--
Flagellation Museum and Library	Jerusalem 28-29-36
Le Sexy	Paris 225-25-17
Drink	Rio de Janeiro 257-7068
Amir's Palace	Doha 25241
Tamil United Liberation Front	Jaffna 7176
Konstantin Chernenko	Moscow 206-25-11
Sun Myong Moon (The Reverend)	New York 730-5782
The Fire Baptised Holiness Church	Independence 331-3049
The Sultan of Johore	Johore JB 2000
Professional Rodeo Cowboys Assocn	Denver 455-3270
World Peace Council	Helsinki 640004

SOME SUMS!

$$1 \times 8 + 1 = 9$$
$$12 \times 8 + 2 = 98$$
$$123 \times 8 + 3 = 987$$
$$1234 \times 8 + 4 = 9876$$
$$12345 \times 8 + 5 = 98765$$
$$123456 \times 8 + 6 = 987654$$
$$1234567 \times 8 + 7 = 9876543$$
$$12345678 \times 8 + 8 = 98765432$$
$$123456789 \times 8 + 9 = 987654321$$

$$9 \times 9 + 7 = 88$$
$$98 \times 9 + 6 = 888$$
$$987 \times 9 + 5 = 8888$$
$$9876 \times 9 + 4 = 88888$$
$$98765 \times 9 + 3 = 888888$$
$$987654 \times 9 + 2 = 8888888$$
$$9876543 \times 9 + 1 = 88888888$$
$$98765432 \times 9 + 0 = 888888888$$

65359477124183 × 17 × 1 = 1111111111111111
65359477124183 × 17 × 2 =2222222222222222
65359477124183 × 17 × 3 = 3333333333333333
65359477124183 × 17 × 4 = 4444444444444444
65359477124183 × 17 × 5 = 5555555555555555
65359477124183 × 17 × 6 = 6666666666666666
65359477124183 × 17 × 7 = 7777777777777777
65359477124183 × 17 × 8 = 8888888888888888
65359477124183 × 17 × 9 = 9999999999999999

1 × 9 − 1 = 8
21 × 9 − 1 = 188
321 × 9 − 1 = 2888
4321 × 9 − 1 = 38888
54321 × 9 − 1 = 488888
654321 × 9 − 1 = 5888888
7654321 × 9 − 1 = 68888888
87654321 × 9 − 1 = 788888888
987654321 × 9 − 1 = 8888888888

1 × 9 + 2 = 11
12 × 9 + 3 = 111
123 × 9 + 4 = 1111
1234 × 9 + 5 = 11111
12345 × 9 + 6 = 111111
123456 × 9 + 7 = 1111111
1234567 × 9 + 8 = 11111111
12345678 × 9 + 9 = 111111111

$111 \quad 1 + 1 + 1 = 3$ $111 \div 3 = 37$

$222 \quad 2 + 2 + 2 = 6$ $222 \div 6 = 37$

$333 \quad 3 + 3 + 3 = 9$ $333 \div 9 = 37$

$444 \quad 4 + 4 + 4 = 12$ $444 \div 12 = 37$

$555 \quad 5 + 5 + 5 = 15$ $555 \div 15 = 37$

$666 \quad 6 + 6 + 6 = 18$ $666 \div 18 = 37$

$777 \quad 7 + 7 + 7 = 21$ $777 \div 21 = 37$

$888 \quad 8 + 8 + 8 = 24$ $888 \div 24 = 37$

$999 \quad 9 + 9 + 9 = 27$ $999 \div 27 = 37$

$$11^2 = 121$$
$$111^2 = 12321$$
$$1111^2 = 1234321$$
$$11111^2 = 123454321$$
$$111111^2 = 12345654321$$
$$1111111^2 = 1234567654321$$
$$11111111^2 = 123456787654321$$
$$111111111^2 = 12345678987654321$$

$$33 \times 3367 = 111111$$
$$66 \times 3367 = 222222$$
$$99 \times 3367 = 333333$$
$$132 \times 3367 = 444444$$
$$165 \times 3367 = 555555$$
$$198 \times 3367 = 666666$$
$$231 \times 3367 = 777777$$
$$264 \times 3367 = 888888$$
$$297 \times 3367 = 999999$$

```
  1 2 3 4 5 6 7 8 9          9 8 7 6 5 4 3 2 1
  1 2 3 4 5 6 7 8              8 7 6 5 4 3 2 1
  1 2 3 4 5 6 7                  7 6 5 4 3 2 1
  1 2 3 4 5 6                      6 5 4 3 2 1
  1 2 3 4 5                          5 4 3 2 1
  1 2 3 4                              4 3 2 1
  1 2 3                                  3 2 1
  1 2                                      2 1
+ 1                                          1
───────────────────          ───────────────────
1 0 8 3 6 7 6 2 6 9          1 0 8 3 6 7 6 2 6 9
```

T HAT'S LIFE

THE HUMAN BODY

The number of bones in the human body

Skull	22	
Ears	6	
Vertebrae	26	
Ribs	24	
Sternum	4	
Pectoral Girdle	4	
Arms	60	(30 for each)
Hip bones	2	
Legs	58	(29 each)
Total	206	

Pulse rates

Age	Pulse rate
Birth	135 beats per minute
2 years	110 beats per minute
4 years	105 beats per minute
6 years	95 beats per minute
8 years	90 beats per minute
10 years	87 beats per minute
15 years	83 beats per minute
20 years	71 beats per minute
21-25 years	74 beats per minute
25-30 years	72 beats per minute
35-40 years	70 beats per minute

40-45 years	72 beats per minute
45-50 years	72 beats per minute
55-60 years	75 beats per minute
60-65 years	73 beats per minute
65-70 years	75 beats per minute
70-75 years	75 beats per minute
75-80 years	72 beats per minute
80+ years	78 beats per minute

YOU COULD MAKE. . .

. . .seven bars of soap with all the fat in the adult human body.

. . .an iron nail three inches long with all the iron in the adult human body.

. . .the lead for 900 pencils from all the carbon in the adult human body.

. . .the heads of 2,000 matches from all the phosphorus in the adult human body.

. . .enough whitewash to paint a small chicken-coop with all the lime in the adult human body.

WHAT HAVE WE HERE?

This description was written by Samuel Wilberforce, Bishop of Oxford, over a century ago:

I have a large Box, with two lids, two caps, three established Measures, and a great number of articles a Carpenter cannot do without. – Then I have always by me a couple of good Fish, and a number of a smaller tribe, – besides two lofty Trees, fine Flowers, and the fruit of an indigenous Plant; a handsome Stag; two playful animals; and a number of smaller and less tame Herd:– Also two Halls, or Places of Worship; some Weapons of warfare; and many Weathercocks: – The Steps of an Hotel; The

House of Commons on the eve of a Dissolution; Two Students or Scholars, and some Spanish Grandees, to wait upon me.

All pronounce me a wonderful piece of Mechanism, but few have numbered up the strange medley of things which compose my whole.

What *exactly* was the Bishop describing?

(If you don't know the answer, Lewis Carroll did and you will find it at the back.)

UNIVERSAL NUMERALS

Egyptian Numerals

While the first method of counting was almost certainly by using the fingers and thumbs on both hands, as long ago as 3000 BC symbols to represent numbers had been invented in Egypt, Babylon, China and India, and by 2900 BC the Egyptians had begun to use numbers in a practical way.

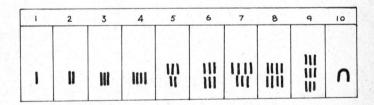

Arabic Numerals

The numerals we use today are Arabic and came to Europe around AD 1200. The chart on page 136 shows how they developed.

	1	2	3	4	5	6	7	8	9	0
1	ᘔ	⅄	୶	Ս	(౨)	⁓	Ч	ﻠ		
2	I	ટ	ૐ	∅	ﺵ	౨	∖	1	૧	O
3	I	ટ	ૐ	૪	Ч	৬	7	8	૧	
4	I	૮	૱	૮	૫	6	7	8	૧	O
5	)	૬	૪	૬	ﺍ	౨	∨	∧	૧	.
6	I	૬	૪	૪	△	౨	∨	∧	૧	.
7	I	2	૪	૪	8	Ч	∨	∧	૧	૮
8	I	૬	૪	૬	౦	7	∨̈	∧	૧	.
9	I	૬	↺	૱	૫	౨	⅄	8	૧	
10	I	૮	૱	ￍ	૫	૫	∧	8	૧	
11	I	૬	Ꮧ	B	૫	౼	∧	૪	౧	
12	I	૬	૱	ￍ	૫	౼	∧	8	૧	⊙
13	I	૮	÷	ￍ	૫	౼	∕	8	૧	
14	૧	૧	૧	૧	૧	1	⅃	⅃	√	
15	I	૮	ￌ	ￍ	8	Ч	∨	∧	૧	
16	I	૮	ￌ	ꜱ	౦	Ч	∨	∧	૧	.
17	I	2	3	૪	૧	6	∨	8	9	૦
18	I	૱	ￌ	5	B	૮	∨	૪	૧	૦
19	I	2	3	4	5	6	7	8	9	૦
20	1	2	3	4	5	6	7	8	9	0

1 Devanagari letters of the 2nd century AD.
2 Arabic numerals of the tenth century.
3 The earliest examples of Arabic numerals in Latin manuscripts, Escorial Library, Spain, 976.
4 Other forms of the Arabic numerals, Western type.
5-8 Arabic numerals, Eastern type (eight modern Arabic numerals as employed in Arabic script).
9-13 The so-called apices of Boethius of the eleventh and twelfth centuries.
14 The numerals of John Basingestokes (d.1252).

136

15-17 Arabic-Byzantine numerals of the twelfth to the fifteenth centuries.
18 Numerals in a manuscript from France (now in Berlin), of the second half of the twelfth century.
19 Numerals in an Italian manuscript from Florence of the first half of the fourteenth century.
20 Numerals in an Italian manuscript of the fifteenth century.

Hebrew Numerals (using letters of the alphabet)

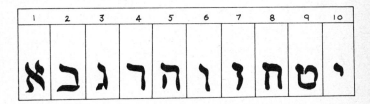

Mayan Numerals

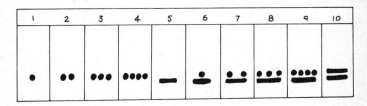

Ionic Numerals (ancient Greek alphabet)

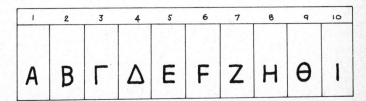

Babylonian Numerals

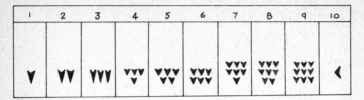

1	2	3	4	5	6	7	8	9	10

Hindu Numerals

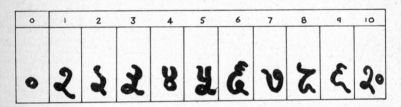

0	1	2	3	4	5	6	7	8	9	10

Chinese Numerals

1	2	3	4	5	6	7	8	9	10

Chinese Rod Numerals

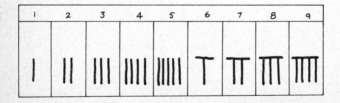

1	2	3	4	5	6	7	8	9

138

Semaphore Numerals

The letters A to K (excluding J) of the semaphore alphabet provide the numerals from one to zero, like this:

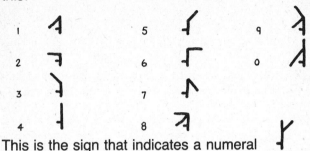

This is the sign that indicates a numeral

Morse Code Numerals

The numerals in Morse Code follow a regular pattern, like this:

1	.----	5	.....	9	----.
2	..---	6	-....	0	-----
3	...--	7	--...		
4	....-	8	---..		

International Numeral Flags

This International Code enables ships to communicate, even if the crews speak different languages. There are ten pennants for the numerals from one to zero, which look like this:

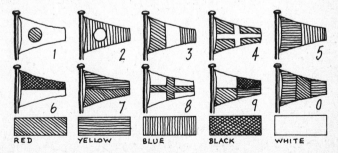

RED   YELLOW   BLUE   BLACK   WHITE

139

Counting by Hand

For thousands of years people have been counting on their fingers – and not just up to ten. Back in the fourth century BC Herodotus referred to the Greek system that enabled you to count on your hands from one to numbers in the thousands. The Romans had a finger-counting system too. So did the ancient Chinese.

The system shown here is one described by a late fifteenth century Italian friar, Luca Pacioli. Many of the symbols had probably been in use for over a thousand years. Writing in the fourth century, St Jerome mentioned the association of marriage with the number 30, while he connected 60 with widowhood. If you look at the two symbols in the chart, you'll see that 30 is circle formed with the thumb and first finger while 60 shows the thumb and finger crossed.

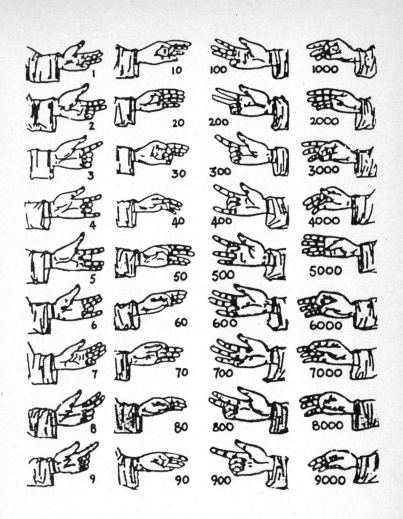

VULGAR FRACTIONS

CENTI-FRACTIONS

Believe it or not, like it or not, each of these horrendous looking fractions and sums equals 100.

a) $3\,\dfrac{69258}{714}$

b) $\dfrac{148}{2.96} + \dfrac{35}{0.7}$

c) $96\,\dfrac{2148}{537}$

d) $70 + 24\,\dfrac{9}{18} + 5\,\dfrac{3}{6}$

e) $82\,\dfrac{3546}{197}$

f) $50 + \dfrac{38}{76} + 49\,\dfrac{1}{2}$

g) $91\,\dfrac{5742}{638}$

h) $80\,\dfrac{27}{54} + 19\,\dfrac{3}{6}$

i) 94 $\dfrac{1578}{263}$

j) 99 $\dfrac{99}{99}$

k) 81 $\dfrac{7524}{396}$

l) 87 + 9 $\dfrac{4}{5}$ + 3 $\dfrac{12}{60}$

m) 96 $\dfrac{1752}{438}$

n) 99 $\dfrac{9}{9}$

o) 81 $\dfrac{5643}{297}$

p) 91 $\dfrac{5823}{647}$

q) 81 $\dfrac{7524}{396}$

MULTIPLICATION = ADDITION

That may sound like a contradiction in terms, but it isn't. Look:

a) $\quad 5 + 1\frac{1}{4} = 6\frac{1}{4}$
and $5 \times 1\frac{1}{4} = 6\frac{1}{4}$

b) $\quad 3 + 1\frac{1}{2} = 4\frac{1}{2}$
and $3 \times 1\frac{1}{2} = 4\frac{1}{2}$

c) $\quad 4 + 1\frac{1}{3} = 5\frac{1}{3}$
and $4 \times 1\frac{1}{3} = 5\frac{1}{3}$

There is a rule that governs the behaviour of these numbers. You will always get matching results from both adding and multiplying if you add or multiply a number by one with a fraction that has one as the numerator (top digit) and has a denominator (bottom digit/s) that is one less than the multiplier. So the same rule would be true of:

$$6 + 1\tfrac{1}{5} = 7\tfrac{1}{5}$$
$$\text{and } 6 \times 1\tfrac{1}{5} = 7\tfrac{1}{5}$$

and

$$9 + 1\tfrac{1}{8} = 10\tfrac{1}{8}$$
$$\text{and } 9 \times 1\tfrac{1}{8} = 10\tfrac{1}{8}$$

For a variation on the same theme, look what you can do with subtraction and division too:

$$\frac{3}{5} - \frac{3}{8} = \frac{9}{40} \qquad \text{and } \frac{3}{5} \times \frac{3}{8} = \frac{9}{40}$$

$$\frac{169}{30} + \frac{13}{15} = \frac{13}{2} \qquad \text{and } \frac{169}{30} \div \frac{13}{15} = \frac{13}{2}$$

$$\frac{121}{28} - \frac{11}{7} = \frac{11}{4} \qquad \text{and } \frac{121}{28} \div \frac{11}{7} = \frac{11}{4}$$

WORDS AND NUMBERS

ARS MAGNA

Under the guidance of one of Europe's most eminent verbivores and numberholics, Mr Darryl Francis, I have been compiling a list of anagrams of all the number names from ONE to TWENTY, and every tenth number name from there to ONE HUNDRED. One number name in this series remains unanagrammable. We searched for possible words such as GITHEY, YEIGHT and YIGHTE, but all to no avail. Another number name, FIVE, yielded an anagram which is not English but Old Norse! Although this goes beyond the bounds of acceptability, it seemed better to include some anagram, in any language at all, rather than to admit defeat.

For the first ten number names, I have listed every possible anagram that I found; for the other number names, I have confined myself to the best.

1   There are six ways in which the letters O, N and E can be arranged, and all of these happen to be words. The commonest of the five anagrams of ONE is EON. Additionally, a NEO is an advocate of that which is novel; NOE is a Biblical proper name; ENO' is a seventeenth century spelling of

'enough'; and OEN is an obsolete spelling of 'owe'.

2   The commonest anagram is TOW, of course. But both OWT, a dialect English variant of 'aught' and 'ought', and WOT, a verb meaning to know or have knowledge of, exist.

3   Of the nine anagrams I was able to find here (the greatest collection for any number name), two, ETHER and THERE, are very common words, not requiring any kind of definition. The other seven anagrams are defined here:

ERTHE   an obsolete variant spelling of 'earth'
HERTE   an obsolete form of 'hart', 'heart' and 'hurt'
HETER   an adjective meaning severe or cruel
REHET   entertainment
REHTE   a thirteenth century form of 'reached'
RETHE   an adjective meaning severe or cruel
THEER   a fifteenth century form of 'there'

Note that HETER and RETHE have exactly the same meaning!

4   The only anagram I was able to track down was ROUF, an obsolete form of both 'roof' and 'rough'

5   From Old Norse comes VEIF, a flapping or waving thing.

6   The only anagram here is XIS, the plural of XI, the fourteenth letter of the Greek alphabet.

7   The commonest anagram is EVENS. However, there are at least four other anagrams. NEVES are types of granular snow; VENES is an obsolete form of 'veins'; EEVNS is a seventeenth century form of 'evens'; and VESEN is a thirteenth century variant of the verb 'freeze', to drive off, to impel, to flog.

8   TEIGH is an obsolete word, the past tense of the

verb 'tee', meaning to draw.

9 There are two anagrams here, both obsolete forms of everyday words. INNE is an obsolete variant of 'in' and 'inn'; and NEIN is a fourteenth century form of 'nine'. You will probably also know that NEIN is German for 'No'.

10 NET springs to mind at once. A second anagram is ENT, a scion or graft, and also a metaphysical term for an existent unity.

11 LEVENE is a surname which can be found in various biographical dictionaries and telephone directories. For example, Aaron Theodorus Levene was a Russian-born American chemist who lived from 1869 until 1940.

12 VELWET is a fifteenth and sixteenth century form of the common word 'velvet'. Interestingly, WELVET is another old spelling of the same word!

13 THREITEN is a scots form of 'threaten'.

14 NEETROUF is a slang term for the number 'fourteen'. It is formed by spelling 'fourteen' backwards, but with the 'ou' left unreversed.

15 FIFTENE is a spelling of 'fifteen' which existed between the thirteenth and sixteenth centuries.

16 SEXTINE is a type of poem.

17 SEVENTENE is a Middle English form of 'seventeen'.

18 TEHEEING is the present participle of the verb 'tehee', meaning to titter.

19 NINETENE is merely a fourteenth century spelling of 'nineteen'.

20 Just one anagram came to light here: TWYNTE, a variant spelling of 'twynt', a rare obsolete noun meaning a jot or particle.

30 Of all the anagrams TYRITH was the most difficult to unearth. Basically, this is the -eth form of the verb 'tyre', a fourteenth to seventeenth century form of 'tire', meaning to tear flesh with the beak. TYRITH appears in an obscure quotation in The Oxford English Dictionary: 'An hawke tyrith vppon Rumppys, she fedith on all manner of flesh'. This quotation goes back to the year 1486.

40 TROFY is a suggested reformed spelling of 'trophy'. At one time when spelling reform was the vogue, many dictionaries actually included reformed spellings of many common words. TROFY was one such.

50 TIFFY is an adjective meaning peevish.

60 At first sight, one might think this number name was incapable of being anagrammed. However, the immensely useful word XYSTI, the plural of 'xystus', is shown in various dictionaries. XYSTI are tree-lined walks.

70 SEVYNTE is a fourteenth century, north of England form of 'seventh'

80 So far this number name has flummoxed both me and Darryl Francis. But an anagram must exist somewhere. There are, after all 719 different ways in which the letters of EIGHTY can be rearranged. One of them must be a word!

90 The only anagram found was TYNNIE, a sixteenth century form of 'tinny'.

100 Although I was unable to find a genuine anagram for this number name, Darryl Francis did manage to create the coinage UNDERHONED. Though this doesn't appear in any dictionary, its meaning is obvious: not honed or sharpened sufficiently for the purpose required.

That Darryl Francis speaks perfect English is not surprising. He's British after all. But because he *is* British it is doubly surprising that he speaks such good French, German, Italian and Spanish as well. He has been looking at the number names from ONE to TWENTY in those languages and trying to make anagrams of them in English: e.g. getting CONIC from CINCO, the Spanish for five, and RIDE from DREI, the German for three. Here are some of his other efforts, going from one (almost) to twenty.

(The letters F, G, I and S merely denote French, German, Italian and Spanish.)

1	G	eins/sine
2	S	dos/sod
3	F	trois/riots
4	S	cuatro/turaco
5	I	cinque/quince
6	G	sechs/chess
7	F	sept/pest
8	G	acht/chat
9	I	nove/oven
10	S	diez/zide
11	F	onze/zone
12	S	doce/code
13	S	trece/erect
14		
15	F	quinze/zequin
16	F	seize/sieze
17		
18		
19	F	dix-neuf/unfixed
20	I	venti/vinet

All the English words appear in Webster's Third New International Dictionary, except for ZIDE (a dialect form of 'side' and SIEZE (another spelling of 'seize'), which are both in The Oxford English Dictionary. Perhaps you have a knowledge of other languages and can fill the gaps at 14, 17 and 18, or even beyond 20. If so, please let me know.

SPYCRAFT

Some people – like Darryl Francis – have a genius for turning numbers into words. Others have a facility for turning words into numbers.

In the top secret world of international espionage the use of numbers as substitutes for letters of the alphabet is a common practice. The most obvious system is to number the letters from one to 26, starting with A, but even the most naive agent is usually able to come up with something more sophisticated than that. For example, here are three simple cipher sequences that were actually used during the First World War.

	(a)	(b)	(c)
A	31	91	146
B	32	92	145
C	33	93	144
D	34	94	143
E	35	95	142
F	36	96	141
G	37	97	140
H	38	98	139
I	39	99	138
J	40	100	137
K	41	101	136
L	42	102	135
M	43	103	134
N	44	104	133
O	45	105	132
P	46	106	131
Q	47	107	130
R	48	108	129
S	49	109	128
T	50	110	127
U	51	111	126
V	52	112	125
W	53	113	124

	(a)	(b)	(c)
X	54	114	123
Y	55	115	122
Z	56	116	121

Obviously it doesn't matter how the numbers and letters are paired, as long as there is a sequence of 26 to cover the whole alphabet. As a high-powered and highly numerate agent, once you've decided on your system, your next move is to select the letters you need to send your message and then write them out in a continuous row. Let's say that you're sending the message:

OH JOY OF NUMBERS using the code under a)

This is what you do:
a) Allocate numbers to the letters in the message.
b) Write the numbers in one continuous row.
c) Divide them into groups, at random which looks like this:

a)
OH	JOY	OF	NUMBERS
45/38/	40/45/55	45/36	44/51/43/32/35/48/49

b) 45384045554536445143332354849

c) 4 5384 04 55 545364 45143 3 235 484 9

Providing the decipherer knows the basis on which the letters and the numbers were paired, all that has to be done is for the digits to be paired again from their random row and then for the letters to be extracted from them.

Mirabeau Cipher

To make your cipher somewhat harder to crack, you may care to introduce a keyword which will enable you to jumble the letters of the alpabet before numbering them. In the case of the famous Mirabeau Cipher, the letters in each group are given a number in numerical

order. Let's suppose that your key word is CRYP-
TOGRAPHY, this is how you might arrange the code:

6	3	8	1	4	7
CRYPT	OGRAP	HYBDE	FIJKL	MNQSU	VWXZ
12345	12345	12345	12345	12345	1234

So to send the message NEED MORE MONEY, you
would produce a row of numbers with two digits
corresponding to each of the letters in the message.
The first digit tells the decipherer which group of letters
is in, the second digit refers to the bottom line of digits
and indicates which letter is meant. Your message in
this instance will read   :

   4285858441 3162854131428563

and it could be transmitted in any order, like this
perhaps:

   428 5 85844 131 6285 413 2 856 3

Fractions Cipher

With this next number cipher, the principle is the same
as for Mirabeau, but the message is transmitted as a
series of *fractions*. Once again a keyword is used to
scramble the alphabet, although this time the bottom
row of digits can be arranged in any order.
For example, this is how the keywords LEXICON could
be used to send a message using the fraction cipher:

1	2	3	4	5	6	7	8
LEX	ICON	ABD	FGHJK	MPQ	RSTUV	WXY	Z
379	4689	789	56789	579	56789	689	9

And this is the message:

2421	7726	2261	3711	1322	1
4997	6988	6883	9673	3989	7

Can you work out what it is? (Answer at the back.)

Twisnum Cipher

With this system, the 26 letters of the alphabet and the ten digits from zero to nine are indicated by pairs of numbers that can be used either way round. This provides for greater security, since each letter has two cipher equivalents. Here is what the code would look like with the alphabet scrambled by the word THURSDAY.

T	12	21
H	13	31
U	14	41
R	15	51
S	16	61
D	17	71
A	18	81
Y	19	91
B	23	32
C	24	42
E	25	52
F	26	62
G	27	72
I	28	82
J	29	92
K	34	43
L	35	53
M	36	63
N	37	73
O	38	83
P	39	93
Q	45	54
V	46	64
W	47	74
X	48	84
Z	49	94

And here are the digits from zero to nine:

1	56	65
2	57	75
3	58	85
4	59	95
5	67	76
6	68	86
7	69	96
8	78	87
9	79	97
0	89	98

Can you decipher this message, remembering that the ciphers can be given in two forms:

3682   6713   5215   2528   4238   6325

(Answer at the back.)

Map Reference Ciphers

This system works on the principle of placing the 26 letters of the alphabet and the digits from one to nine inside a square, with the cipher numbers on the outside, along the top and down the left-hand side. Each letter is then sent in the code as a cross-reference. All the decipherer has to do is to enter the coded numbers against his own cipher square to be able to extract the message. Here's the basic alphabet in a cipher square, but, as always, the alphabet could be scrambled by the use of a keyword:

	0	1	2	3	4	5	6
7	A	B	C	D	E	F	G
3	H	I	J	K	L	M	N
9	O	P	Q	R	S	T	U
4	V	W	X	Y	Z	1	2
2	3	4	5	6	7	8	9

Now as a final test of your potential as a masterspy,

see if you can send these messages:

a) HELP BRAIN IS CRACKING UP
b) THANK YOU MINE IS A PINT
c) NEVER ON A SUNDAY

(Answers at the back.)

The Cipher-Breaker's Treasury
I trust this book won't fall into the hands of enemy agents. On the assumption that you're an English speaker and on *our side*, let me reveal to you the relative frequencies of different letters of the alphabet as used in written English. This information will be invaluable to you if ever you are called upon to crack a number cipher.

1 Letters of highest frequency in descending order are:
E, T, A, O, N, I, R, S, H (These account for about 70 per cent of a given text.)

2 Letters of medium frequency are (in descending order):
D, L, U, C, M.

3 Letters of low frequency are (in descending order):
P, F, Y, W, G, B, V.

4 Letters of the lowest frequency (in descending order):
K, X, Q, J, Z.

5 Vowels will account for about 40 per cent of the letters in a given text.

6 Half of the words in English begin with:
A, O, S, T, W.

7 Half of all the words in English end with:
D, E, S, T.

8 In English the most frequently used double letters are:
EE, FF, LL, OO, SS, TT.

9 Other common doubles are:
CC, MM, NN, PP, RR.

10 Using a basis of 10,000 word counts, the ten commonest words in English, arranged by letter totals, are, in order of frequency:

two-letter: of, to, in, it, is, be, as, at, so, we.

three-letter: the, and, for, are, but, not, you, all, any, can.

four-letter: that, with, have, this, will, your, from, they, know, want.

X = 10

ROMAN NUMERALS

Here are the principle Roman numerals from one to 1,000,000. The line above some of the numerals is a device for raising the value of the numeral in question by 1,000.

1	I	500	D	or	IƆ
2	II	600	DC	or	IƆC
3	III	700	DCC	or	IƆCC
4	IV or IIII	800	DCCC	or	IƆCCC
5	V	900	CM	or	IƆCD
6	VI	1,000	M	or	CIƆ
7	VII	1,100	MC	or	CIƆC
8	VIII	1,200	MCC	or	CIƆCC
9	IX	1,300	MCCC	or	CIƆCCC
10	X	1,400	MCD	or	CIƆCD
11	XI	1,500	MD	or	CIƆD
12	XII	1,600	MDC	or	CIƆDC
13	XIII	1,700	MDCC	or	CIƆDCC
14	XIV	1,800	MDCCC	or	CIƆDCCC
15	XV	1,900	MCM	or	CIƆIƆCD
16	XVI	2,000	MM	or	CIƆCI
17	XVII	3,000	MMM		
18	XVIII	4,000	$\overline{IV}$		
19	XIX	5,000	$\overline{V}$	or	IƆƆ
20	XX	6,000	$\overline{VI}$		
30	XXX	7,000	$\overline{VII}$		
40	XL	8,000	$\overline{VIII}$		
50	L	9,000	$\overline{IX}$		
60	LX	10,000	$\overline{X}$	or	CCƆƆ
70	LXX	50,000	$\overline{L}$	or	IƆƆƆ
80	LXXX	100,000	$\overline{C}$	or	CCCIƆƆƆ
90	XC	500,000	$\overline{D}$	or	IƆƆƆƆ
100	C	1,000,000	$\overline{M}$	or	CCCCIƆƆƆƆ
200	CC				
300	CCC				
400	CD or CCCC				

YEAR IN YEAR OUT

DAY BY DAY

How's your day divided? Breakfast; lunch; tea; supper? Catch the 8.05; 11 o'clock cuppa; lunch hour 12.30 to 2.00; 4.00 o'clock cuppa; catch the 5.41? We all look to the day in different ways according to our particular way of life. This is how the Antananarivo of Madagascar divided their day, starting with midnight:

Halving-of-the-night
Frog-crowing
Cock-crowing
Morning-and-night-together
Crow-crowing
Bright-horizon
Glimmer-of-the-day
Colours-of-cattle-can-be-seen
Sunrise
Broad-day
Dew-falls
Cattle-go-out
Leaves-dry
Hoar-frost-disappears
Sun-over-the-ridge-of-roof
Sun-enters-threshold-of-room

Sun-at-rice-pounding-place
Sun-at-house-post
Sun-at-sheep-pen
Cattle-come-home
Sunset-flush
Sun-dead
Fowls-come-in
People-cook-rice
People-eat-rice
People-finish-eating
Go-to-sleep

It makes our digital clocks, time-checks, wrist-watches and timetables seem somewhat prosaic, doesn't it?

Our Saxon forebears, on the other hand, moved a stage nearer our digital deliniation of the day by using sun-dials, though only four of their actual divisions (tides) could be recorded in daylight. Here are the 'tides' of the Saxon day:

4.30 am – 7.30 am	Morgan
7.30 am – 10.30 am	Dael-mael
10.30 am – 1.30 pm	Mid-daeg
1.30 pm – 4.30 pm	Ofanverthrdagr
4.30 pm – 7.30 pm	Mid-aften
7.30 pm – 10.30 pm	Ondverthnott
10.30 pm – 1.30 am	Mid-niht
1.30 am – 4.30 am	Ofanverthnott

And coming right up-to-date, though using a day division that has been around for centuries now, this is how the watches are spread at sea:

Midnight – 4 am	Middle Watch
4 am – 8 am	Morning Watch
8 am – Noon	Forenoon Watch
Noon – 4 pm	Afternoon Watch
4 pm – 6 pm	First Dog Watch
6 pm – 8 pm	Second Dog Watch

| 8 pm – Midnight | First Watch |

(The purpose of the two short (dog) watches is to prevent the same men from being on duty for the same period each day.)

MONTH BY MONTH

The Romans had ten months in their year until Numa Pompilius decided it was time to add a couple more. For centuries in Bali, Samoa, parts of East Africa and New Guinea the year consisted of six months, and until not long ago the Dayak of Borneo reckoned their year according to the eight different labours they performed in the field.

In Australia the Bigambul tribe used to divide their year into four seasons in accordance with the blossoming of certain trees:

Autumn: (Yerrabinda) The Yerra tree blossoms
Winter: (Nigabinda) The Niga tree blossoms
Spring: (Wobinda)   The Iron-bark tree blossoms
Summer: (Ground   No tree blossoms
burns feet)

The Maoris of New Zealand used to work on the basis of 13 months to the year:

1 Month of the Great White Star.
2 Month of another star.
3 Spring begins.
4 Tree flowering month.
5 Cuckoo arrives.
6 Rewarewa flowers.
7 Nga flowers.
8 Great summer star.
9 Dry and scarce month.

10 Harvest.
11 Cuckoo leaves.
12 Winter star.
13 Grumbling month.

The Chinese developed a system of dividing their year into units of approximately 15 days to give no fewer than 24 seasons:

     1 Season of rainwater (15 days).
     2 Season of moving snakes (15 days).
     3 Spring equinox (15 days).
     4 Season of pure brightness (15 days).
5 & 6 Season of sowing rain and season of dawn summer (31 days).
7 & 8 Season of little fruitfulness and season of corn in the beard (31 days).
     9 Summer solstice (16 days).
  10 Season of beginning of heat (16 days).
11 & 12 Season of great heat and season of signs of autumn (31 days).
13 & 14 Season of end of heat and season of white dew (31 days).
  15 Season of cold dew (15 days).
  16 Autumn equinox (15 days).
  17 Season of hoar frost (15 days).
  18 Season of signs of winter (15 days).
19 & 20 Season of beginning of snow and season of Great Snow (29 days).
  21 Winter solstice (15 days).
  22 Season of little cold (15 days).
  23 Season of great cold (15 days).
  24 Season of dawn of spring (15 days).

CALENDARS FOR TIDY MINDS

Nowadays, throughout the world, most people have

come to accept that some years have 365 days in them while others have 366, and that some months contain 31 days while others run to only 30 and one must made do with either 28 or 29. The system might seem eccentric to a creature from Mars, but we earthlings have got used to it and are happy to muddle along as we are.

Occasionally an individual with an exceptionally tidy mind comes onto the scene and does his best to straighten us out. One such was the French philosopher Auguste Comte who devised the International Fixed Calendar back in 1849. The essence of Monsieur Comte's calendar was this:

a) A year divided into 13 months.

b) Each month to have exactly 28 days.

c) The extra month to be squeezed between June and July, at the middle of the year, and to be called Sol.

d) An extra holiday, without regard to month or day to be added at the end of the year (after 28 December).

e) A second extra-calendar holiday to be added every leap year.

f) The first day of each month to be a Sunday.

g) The last day to be a Saturday.

Comte's scheme didn't find as much favour as he had hoped, principally because his year did not divide conveniently into quarters, and the French business community felt that it should.

A development of Comte's scheme was put forward by a fellow Frenchman, Armelin, in 1884. He proposed:

a) Maintaining 12 months.

b) Equalizing the quarters by giving each one 91 days.

c) So, January, April, July and October would all have 31 days.

d) And all other months would have 30 days.

e) The 365-day-year and the leap year provisions were to follow Comte's ideas.

Even though 47 countries took up Armelin's World Calendar in the 1930s, it was again the business community who poured cold water on the idea. This time they argued that each month spanned five different weeks, while each month of a given quarter started on a different day.

More recently American industrial engineer Wallace Barlow has produced another alternative calendar, consisting of 12 months, each of 28 days, with a series of extra-calendar holidays falling at the end of each month (a development of Comte's single extra-calendar holiday at the end of the year). Barlow's calendar would work like this:

a) Each month would last 28 days.

b) Every month would begin on a Monday.

c) Apart from Saturday and Sunday each week, there would be no extra holiday during any month.

d) All festivals and holidays would be collected at the end of each month (there would be 29 of these).

This would have the added effect of shortening the working week to 38 hours, which in turn would shorten the working year by 14 days. However, without midweek holidays to interrupt the working schedule, the inventor argues that increased productivity would compensate for this loss.

This is how each month would look in the Barlow Calendar:

M	T	W	T	F	S	S
1	2	3	4	5	6	7
8	9	10	11	12	13	14
15	16	17	18	19	20	21
22	23	24	25	26	27	28

Easter Day would always fall on 28 April, but all the other holidays and festivals would be arranged like this:

January 29th	Winter Festival
February 29th	National Hero's Day
March 29th	Spring Festival
April 29th, 30th and 31st	Easter Holiday
May 29th, 30th, 31st, 32nd (and on Leap Year 33rd)	Summer Festival
June 29th, 30th, 31st and 32nd	Festival of Peace
July 29th, 30th and 31st	Festival of Independence
August 29th and 30th	Festival of Friendship
September 29th and 30th	Autumn Festival
October 29th	Music Festival
November 29th and 30th	Thanksgiving Day
December 29th, 30th, 31st, 32nd and 31st	Christmas Holiday

THANK YOU FOR HAVING ME

Given that you or somebody you know – most probably your admirable local librarian – have been generous enough to pay good money for this book, I rather feel the least I can do is attempt to repay your kindness in some small measure. That's why I'm giving you this chapter.

The chances are you own a calendar. The chances are you paid for it. The chances *were* that you were planning to buy another calendar for next year. And for the following year. And for the year after that. Well, now you don't need to. Between today and the end of the century you won't have to spend another penny on calendars – thanks to Brandreth's Calendar of the Century.

In future if you need a calendar just select the year you're after from the list that follows and turn to the calendar in question which will be one of the 14 featured over the next few pages. And if you want to look back instead of forward you can, because I'm giving you calendars for every year from the first of the century – 1901 – to the last – 2000.

I've always known I was born on the 8th March 1948, but until I created the Calendar of the Century I didn't know it was a Monday. Now you too can discover if you're fair of face or full of grace, full of woe or have far to go, are loving and giving, work hard for your living or are lucky enough to be bonny and blithe and gay.

1901	3	1935	3	1968	9
1902	4	1936	11	1969	4
1903	5	1937	6	1970	5
1904	13	1938	7	1971	6
1905	1	1939	1	1972	14
1906	2	1940	9	1973	2
1907	3	1941	4	1974	3
1908	11	1942	5	1975	4
1909	6	1943	6	1976	12
1910	7	1944	14	1977	7
1911	1	1945	2	1978	1
1912	9	1946	3	1979	2
1913	4	1947	4	1980	10
1914	5	1948	12	1981	5
1915	6	1949	7	1982	6
1916	14	1950	1	1983	7
1917	2	1951	2	1984	8
1918	3	1952	10	1985	3
1919	4	1953	5	1986	4
1920	12	1954	6	1987	5
1921	7	1955	7	1988	13
1922	1	1956	8	1989	1
1923	2	1957	3	1990	2

1924	10	1958	4	1991	3
1925	5	1959	5	1992	11
1926	6	1960	13	1993	6
1927	7	1961	1	1994	7
1928	8	1962	2	1995	1
1929	3	1963	3	1996	9
1930	4	1964	11	1997	4
1931	5	1965	6	1998	5
1932	13	1966	7	1999	6
1933	1	1967	1	2000	14
1934	2				

1

	January	February	March	April
M	2 9 16 23 30	6 13 20 27	6 13 20 27	3 10 17 24
T	3 10 17 24 31	7 14 21 28	7 14 21 28	4 11 18 25
W	4 11 18 25	1 8 15 22	1 8 15 22 29	5 12 19 26
T	5 12 19 26	2 9 16 23	2 9 16 23 30	6 13 20 27
F	6 13 20 27	3 10 17 24	3 10 17 24 31	7 14 21 28
S	7 14 21 28	4 11 18 25	4 11 18 25	1 8 15 22 29
S	1 8 15 22 29	5 12 19 26	5 12 19 26	2 9 16 23 30

	May	June	July	August
M	1 8 15 22 29	5 12 19 26	3 10 17 24 31	7 14 21 28
T	2 9 16 23 30	6 13 20 27	4 11 18 25	1 8 15 22 29
W	3 10 17 24 31	7 14 21 28	5 12 19 26	2 9 16 23 30
T	4 11 18 25	1 8 15 22 29	6 13 20 27	3 10 17 24 31
F	5 12 19 26	2 9 16 23 30	7 14 21 28	4 11 18 25
S	6 13 20 27	3 10 17 24	1 8 15 22 29	5 12 19 26
S	7 14 21 28	4 11 18 25	2 9 16 23 30	6 13 20 27

	September	October	November	December
M	4 11 18 25	2 9 16 23 30	6 13 20 27	4 11 18 25
T	5 12 19 26	3 10 17 24 31	7 14 21 28	5 12 19 26
W	6 13 20 27	4 11 18 25	1 8 15 22 29	6 13 20 27
T	7 14 21 28	5 12 19 26	2 9 16 23 30	7 14 21 28
F	1 8 15 22 29	6 13 20 27	3 10 17 24	1 8 15 22 29
S	2 9 16 23 30	7 14 21 28	4 11 18 25	2 9 16 23 30
S	3 10 17 24	1 8 15 22 29	5 12 19 26	3 10 17 24 31

2

	January	February	March	April
M	1 8 15 22 29	5 12 19 26	5 12 19 26	2 9 16 23 30
T	2 9 16 23 30	6 13 20 27	6 13 20 27	3 10 17 24
W	3 10 17 24 31	7 14 21 28	7 14 21 28	4 11 18 25
T	4 11 18 25	1 8 15 22	1 8 15 22 29	5 12 19 26
F	5 12 19 26	2 9 16 23	2 9 16 23 30	6 13 20 27
S	6 13 20 27	3 10 17 24	3 10 17 24 31	7 14 21 28
S	7 14 21 28	4 11 18 25	4 11 18 25	1 8 15 22 29

	May	June	July	August
M	7 14 21 28	4 11 18 25	2 9 16 23 30	6 13 20 27
T	1 8 15 22 29	5 12 19 26	3 10 17 24 31	7 14 21 28
W	2 9 16 23 30	6 13 20 27	4 11 18 25	1 8 15 22 29
T	3 10 17 24 31	7 14 21 28	5 12 19 26	2 9 16 23 30
F	4 11 18 25	1 8 15 22 29	6 13 20 27	3 10 17 24 31
S	5 12 19 26	2 9 16 23 30	7 14 21 28	4 11 18 25
S	6 13 20 27	3 10 17 24	1 8 15 22 29	5 12 19 26

	September	October	November	December
M	3 10 17 24	1 8 15 22 29	5 12 19 26	3 10 17 24 31
T	4 11 18 25	2 9 16 23 30	6 13 20 27	4 11 18 25
W	5 12 19 26	3 10 17 24 31	7 14 21 28	5 12 19 26
T	6 13 20 27	4 11 18 25	1 8 15 22 29	6 13 20 27
F	7 14 21 28	5 12 19 26	2 9 16 23 30	7 14 21 28
S	1 8 15 22 29	6 13 20 27	3 10 17 24	1 8 15 22 29
S	2 9 16 23 30	7 14 21 28	4 11 18 25	2 9 16 23 30

3

	January	February	March	April
M	7 14 21 28	4 11 18 25	4 11 18 25	1 8 15 22 29
T	1 8 15 22 29	5 12 19 26	5 12 19 26	2 9 16 23 30
W	2 9 16 23 30	6 13 20 27	6 13 20 27	3 10 17 24
T	3 10 17 24 31	7 14 21 28	7 14 21 28	4 11 18 25
F	4 11 18 25	1 8 15 22	1 8 15 22 29	5 12 19 26
S	5 12 19 26	2 9 16 23	2 9 16 23 30	6 13 20 27
S	6 13 20 27	3 10 17 24	3 10 17 24 31	7 14 21 28

	May	June	July	August
M	6 13 20 27	3 10 17 24	1 8 15 22 29	5 12 19 26
T	7 14 21 28	4 11 18 25	2 9 16 23 30	6 13 20 27
W	1 8 15 22 29	5 12 19 26	3 10 17 24 31	7 14 21 28
T	2 9 16 23 30	6 13 20 27	4 11 18 25	1 8 15 22 29
F	3 10 17 24 31	7 14 21 28	5 12 19 26	2 9 16 23 30
S	4 11 18 25	1 8 15 22 29	6 13 20 27	3 10 17 24 31
S	5 12 19 26	2 9 16 23 30	7 14 21 28	4 11 18 25

	September	October	November	December
M	2 9 16 23 30	7 14 21 28	4 11 18 25	2 9 16 23 30
T	3 10 17 24	1 8 15 22 29	5 12 19 26	3 10 17 24 31
W	4 11 18 25	2 9 16 23 30	6 13 20 27	4 11 18 25
T	5 12 19 26	3 10 17 24 31	7 14 21 28	5 12 19 26
F	6 13 20 27	4 11 18 25	1 8 15 22 29	6 13 20 27
S	7 14 21 28	5 12 19 26	2 9 16 23 30	7 14 21 28
S	1 8 15 22 29	6 13 20 27	3 10 17 24	1 8 15 22 29

4

	January	February	March	April
M	6 13 20 27	3 10 17 24	3 10 17 24 31	7 14 21 28
T	7 14 21 28	4 11 18 25	4 11 18 25	1 8 15 22 29
W	1 8 15 22 29	5 12 19 26	5 12 19 26	2 9 16 23 30
T	2 9 16 23 30	6 13 20 27	6 13 20 27	3 10 17 24
F	3 10 17 24 31	7 14 21 28	7 14 21 28	4 11 18 25
S	4 11 18 25	1 8 15 22	1 8 15 22 29	5 12 19 26
S	5 12 19 26	2 9 16 23	2 9 16 23 30	6 13 20 27

	May	June	July	August
M	5 12 19 26	2 9 16 23 30	7 14 21 28	4 11 18 25
T	6 13 20 27	3 10 17 24	1 8 15 22 29	5 12 19 26
W	7 14 21 28	4 11 18 25	2 9 16 23 30	6 13 20 27
T	1 8 15 22 29	5 12 19 26	3 10 17 24 31	7 14 21 28
F	2 9 16 23 30	6 13 20 27	4 11 18 25	1 8 15 22 29
S	3 10 17 24 31	7 14 21 28	5 12 19 26	2 9 16 23 30
S	4 11 18 25	1 8 15 22 29	6 13 20 27	3 10 17 24 31

	September	October	November	December
M	1 8 15 22 29	6 13 20 27	3 10 17 24	1 8 15 22 29
T	2 9 16 23 30	7 14 21 28	4 11 18 25	2 9 16 23 30
W	3 10 17 24	1 8 15 22 29	5 12 19 26	3 10 17 24 31
T	4 11 18 25	2 9 16 23 30	6 13 20 27	4 11 18 25
F	5 12 19 26	3 10 17 24 31	7 14 21 28	5 12 19 26
S	6 13 20 27	4 11 18 25	1 8 15 22 29	6 13 20 27
S	7 14 21 28	5 12 19 26	2 9 16 23 30	7 14 21 28

5

	January	February	March	April
M	5 12 19 26	2 9 16 23	2 9 16 23 30	6 13 20 27
T	6 13 20 27	3 10 17 24	3 10 17 24 31	7 14 21 28
W	7 14 21 28	4 11 18 25	4 11 18 25	1 8 15 22 29
T	1 8 15 22 29	5 12 19 26	5 12 19 26	2 9 16 23 30
F	2 9 16 23 30	6 13 20 27	6 13 20 27	3 10 17 24
S	3 10 17 24 31	7 14 21 28	7 14 21 28	4 11 18 25
S	4 11 18 25	1 8 15 22	1 8 15 22 29	5 12 19 26

	May	June	July	August
M	4 11 18 25	1 8 15 22 29	6 13 20 27	3 10 17 24 31
T	5 12 19 26	2 9 16 23 30	7 14 21 28	4 11 18 25
W	6 13 20 27	3 10 17 24	1 8 15 22 29	5 12 19 26
T	7 14 21 28	4 11 18 25	2 9 16 23 30	6 13 20 27
F	1 8 15 22 29	5 12 19 26	3 10 17 24 31	7 14 21 28
S	2 9 16 23 30	6 13 20 27	4 11 18 25	1 8 15 22 29
S	3 10 17 24 31	7 14 21 28	5 12 19 26	2 9 16 23 30

	September	October	November	December
M	7 14 21 28	5 12 19 26	2 9 16 23 30	7 14 21 28
T	1 8 15 22 29	6 13 20 27	3 10 17 24	1 8 15 22 29
W	2 9 16 23 30	7 14 21 28	4 11 18 25	2 9 16 23 30
T	3 10 17 24	1 8 15 22 29	5 12 19 26	3 10 17 24 31
F	4 11 18 25	2 9 16 23 30	6 13 20 27	4 11 18 25
S	5 12 19 26	3 10 17 24 31	7 14 21 28	5 12 19 26
S	6 13 20 27	4 11 18 25	1 8 15 22 29	6 13 20 27

6

	January	February	March	April
M	4 11 18 25	1 8 15 22	1 8 15 22 29	5 12 19 26
T	5 12 19 26	2 9 16 23	2 9 16 23 30	6 13 20 27
W	6 13 20 27	3 10 17 24	3 10 17 24 31	7 14 21 28
T	7 14 21 28	4 11 18 25	4 11 18 25	1 8 15 22 29
F	1 8 15 22 29	5 12 19 26	5 12 19 26	2 9 16 23 30
S	2 9 16 23 30	6 13 20 27	6 13 20 27	3 10 17 24
S	3 10 17 24 31	7 14 21 28	7 14 21 28	4 11 18 25

	May	June	July	August
M	3 10 17 24 31	7 14 21 28	5 12 19 26	2 9 16 23 30
T	4 11 18 25	1 8 15 22 29	6 13 20 27	3 10 17 24 31
W	5 12 19 26	2 9 16 23 30	7 14 21 28	4 11 18 25
T	6 13 20 27	3 10 17 24	1 8 15 22 29	5 12 19 26
F	7 14 21 28	4 11 18 25	2 9 16 23 30	6 13 20 27
S	1 8 15 22 29	5 12 19 26	3 10 17 24 31	7 14 21 28
S	2 9 16 23 30	6 13 20 27	4 11 18 25	1 8 15 22 29

	September	October	November	December
M	6 13 20 27	4 11 18 25	1 8 15 22 29	6 13 20 27
T	7 14 21 28	5 12 19 26	2 9 16 23 30	7 14 21 28
W	1 8 15 22 29	6 13 20 27	3 10 17 24	1 8 15 22 29
T	2 9 16 23 30	7 14 21 28	4 11 18 25	2 9 16 23 30
F	3 10 17 24	1 8 15 22 29	5 12 19 26	3 10 17 24 31
S	4 11 18 25	2 9 16 23 30	6 13 20 27	4 11 18 25
S	5 12 19 26	3 10 17 24 31	7 14 21 28	5 12 19 26

7

January
M	3	10	17	24	31
T	4	11	18	25	
W	5	12	19	26	
T	6	13	20	27	
F	7	14	21	28	
S	1	8	15	22	29
S	2	9	16	23	30

February
M		7	14	21	28
T	1	8	15	22	
W	2	9	16	23	
T	3	10	17	24	
F	4	11	18	25	
S	5	12	19	26	
S	6	13	20	27	

March
M		7	14	21	28
T	1	8	15	22	29
W	2	9	16	23	30
T	3	10	17	24	31
F	4	11	18	25	
S	5	12	19	26	
S	6	13	20	27	

April
M		4	11	18	25
T		5	12	19	26
W		6	13	20	27
T		7	14	21	28
F	1	8	15	22	29
S	2	9	16	23	30
S	3	10	17	24	

May
M	2	9	16	23	30
T	3	10	17	24	31
W	4	11	18	25	
T	5	12	19	26	
F	6	13	20	27	
S	7	14	21	28	
S	1	8	15	22	29

June
M		6	13	20	27
T		7	14	21	28
W	1	8	15	22	29
T	2	9	16	23	30
F	3	10	17	24	
S	4	11	18	25	
S	5	12	19	26	

July
M		4	11	18	25
T		5	12	19	26
W		6	13	20	27
T		7	14	21	28
F	1	8	15	22	29
S	2	9	16	23	30
S	3	10	17	24	31

August
M	1	8	15	22	29
T	2	9	16	23	30
W	3	10	17	24	31
T	4	11	18	25	
F	5	12	19	26	
S	6	13	20	27	
S	7	14	21	28	

September
M		5	12	19	26
T		6	13	20	27
W		7	14	21	28
T	1	8	15	22	29
F	2	9	16	23	30
S	3	10	17	24	
S	4	11	18	25	

October
M	3	10	17	24	31
T	4	11	18	25	
W	5	12	19	26	
T	6	13	20	27	
F	7	14	21	28	
S	1	8	15	22	29
S	2	9	16	23	30

November
M		7	14	21	28
T	1	8	15	22	29
W	2	9	16	23	30
T	3	10	17	24	
F	4	11	18	25	
S	5	12	19	26	
S	6	13	20	27	

December
M		5	12	19	26
T		6	13	20	27
W		7	14	21	28
T	1	8	15	22	29
F	2	9	16	23	30
S	3	10	17	24	31
S	4	11	18	25	

8

January
M	2	9	16	23	30
T	3	10	17	24	31
W	4	11	18	25	
T	5	12	19	26	
F	6	13	20	27	
S	7	14	21	28	
S	1	8	15	22	29

February
M		6	13	20	27
T		7	14	21	28
W	1	8	15	22	29
T	2	9	16	23	
F	3	10	17	24	
S	4	11	18	25	
S	5	12	19	26	

March
M		5	12	19	26
T		6	13	20	27
W		7	14	21	28
T	1	8	15	22	29
F	2	9	16	23	30
S	3	10	17	24	31
S	4	11	18	25	

April
M	2	9	16	23	30
T	3	10	17	24	
W	4	11	18	25	
T	5	12	19	26	
F	6	13	20	27	
S	7	14	21	28	
S	1	8	15	22	29

May
M		7	14	21	28
T	1	8	15	22	29
W	2	9	16	23	30
T	3	10	17	24	31
F	4	11	18	25	
S	5	12	19	26	
S	6	13	20	27	

June
M		4	11	18	25
T		5	12	19	26
W		6	13	20	27
T		7	14	21	28
F	1	8	15	22	29
S	2	9	16	23	30
S	3	10	17	24	

July
M	2	9	16	23	30
T	3	10	17	24	31
W	4	11	18	25	
T	5	12	19	26	
F	6	13	20	27	
S	7	14	21	28	
S	1	8	15	22	29

August
M		6	13	20	27
T		7	14	21	28
W	1	8	15	22	29
T	2	9	16	23	30
F	3	10	17	24	31
S	4	11	18	25	
S	5	12	19	26	

September
M	3	10	17	24	
T	4	11	18	25	
W	5	12	19	26	
T	6	13	20	27	
F	7	14	21	28	
S	1	8	15	22	29
S	2	9	16	23	30

October
M	1	8	15	22	29
T	2	9	16	23	30
W	3	10	17	24	31
T	4	11	18	25	
F	5	12	19	26	
S	6	13	20	27	
S	7	14	21	28	

November
M		5	12	19	26
T		6	13	20	27
W		7	14	21	28
T	1	8	15	22	29
F	2	9	16	23	30
S	3	10	17	24	
S	4	11	18	25	

December
M	3	10	17	24	31
T	4	11	18	25	
W	5	12	19	26	
T	6	13	20	27	
F	7	14	21	28	
S	1	8	15	22	29
S	2	9	16	23	30

9

	January	February	March	April
M	1 8 15 22 29	5 12 19 26	4 11 18 25	1 8 15 22 29
T	2 9 16 23 30	6 13 20 27	5 12 19 26	2 9 16 23 30
W	3 10 17 24 31	7 14 21 28	6 13 20 27	3 10 17 24
T	4 11 18 25	1 8 15 22 29	7 14 21 28	4 11 18 25
F	5 12 19 26	2 9 16 23	1 8 15 22 29	5 12 19 26
S	6 13 20 27	3 10 17 24	2 9 16 23 30	6 13 20 27
S	7 14 21 28	4 11 18 25	3 10 17 24 31	7 14 21 28

	May	June	July	August
M	6 13 20 27	3 10 17 24	1 8 15 22 29	5 12 19 26
T	7 14 21 28	4 11 18 25	2 9 16 23 30	6 13 20 27
W	1 8 15 22 29	5 12 19 26	3 10 17 24 31	7 14 21 28
T	2 9 16 23 30	6 13 20 27	4 11 18 25	1 8 15 22 29
F	3 10 17 24 31	7 14 21 28	5 12 19 26	2 9 16 23 30
S	4 11 18 25	1 8 15 22 29	6 13 20 27	3 10 17 24 31
S	5 12 19 26	2 9 16 23 30	7 14 21 28	4 11 18 25

	September	October	November	December
M	2 9 16 23 30	7 14 21 28	4 11 18 25	2 9 16 23 30
T	3 10 17 24	1 8 15 22 29	5 12 19 26	3 10 17 24 31
W	4 11 18 25	2 9 16 23 30	6 13 20 27	4 11 18 25
T	5 12 19 26	3 10 17 24 31	7 14 21 28	5 12 19 26
F	6 13 20 27	4 11 18 25	1 8 15 22 29	6 13 20 27
S	7 14 21 28	5 12 19 26	2 9 16 23 30	7 14 21 28
S	1 8 15 22 29	6 13 20 27	3 10 17 24	1 8 15 22 29

10

	January	February	March	April
M	7 14 21 28	4 11 18 25	3 10 17 24 31	7 14 21 28
T	1 8 15 22 29	5 12 19 26	4 11 18 25	1 8 15 22 29
W	2 9 16 23 30	6 13 20 27	5 12 19 26	2 9 16 23 30
T	3 10 17 24 31	7 14 21 28	6 13 20 27	3 10 17 24
F	4 11 18 25	1 8 15 22 29	7 14 21 28	4 11 18 25
S	5 12 19 26	2 9 16 23	1 8 15 22 29	5 12 19 26
S	6 13 20 27	3 10 17 24	2 9 16 23 30	6 13 20 27

	May	June	July	August
M	5 12 19 26	2 9 16 23 30	7 14 21 28	4 11 18 25
T	6 13 20 27	3 10 17 24	1 8 15 22 29	5 12 19 26
W	7 14 21 28	4 11 18 25	2 9 16 23 30	6 13 20 27
T	1 8 15 22 29	5 12 19 26	3 10 17 24 31	7 14 21 28
F	2 9 16 23 30	6 13 20 27	4 11 18 25	1 8 15 22 29
S	3 10 17 24 31	7 14 21 28	5 12 19 26	2 9 16 23 30
S	4 11 18 25	1 8 15 22 29	6 13 20 27	3 10 17 24 31

	September	October	November	December
M	1 8 15 22 29	6 13 20 27	3 10 17 24	1 8 15 22 29
T	2 9 16 23 30	7 14 21 28	4 11 18 25	2 9 16 23 30
W	3 10 17 24	1 8 15 22 29	5 12 19 26	3 10 17 24 31
T	4 11 18 25	2 9 16 23 30	6 13 20 27	4 11 18 25
F	5 12 19 26	3 10 17 24 31	7 14 21 28	5 12 19 26
S	6 13 20 27	4 11 18 25	1 8 15 22 29	6 13 20 27
S	7 14 21 28	5 12 19 26	2 9 16 23 30	7 14 21 28

11

January
M		6	13	20	27
T		7	14	21	28
W	1	8	15	22	29
T	2	9	16	23	30
F	3	10	17	24	31
S	4	11	18	25	
S	5	12	19	26	

February
M		3	10	17	24
T		4	11	18	25
W		5	12	19	26
T		6	13	20	27
F		7	14	21	28
S	1	8	15	22	29
S	2	9	16	23	

March
M		2	9	16	23	30
T		3	10	17	24	31
W		4	11	18	25
T		5	12	19	26
F		6	13	20	27
S		7	14	21	28
S	1	8	15	22	29

April
M		6	13	20	27
T		7	14	21	28
W	1	8	15	22	29
T	2	9	16	23	30
F	3	10	17	24	
S	4	11	18	25	
S	5	12	19	26	

May
M		4	11	18	25
T		5	12	19	26
W		6	13	20	27
T		7	14	21	28
F	1	8	15	22	29
S	2	9	16	23	30
S	3	10	17	24	31

June
M	1	8	15	22	29
T	2	9	16	23	30
W	3	10	17	24	
T	4	11	18	25	
F	5	12	19	26	
S	6	13	20	27	
S	7	14	21	28	

July
M		6	13	20	27
T		7	14	21	28
W	1	8	15	22	29
T	2	9	16	23	30
F	3	10	17	24	31
S	4	11	18	25	
S	5	12	19	26	

August
M	3	10	17	24	31
T	4	11	18	25	
W	5	12	19	26	
T	6	13	20	27	
F	7	14	21	28	
S	1	8	15	22	29
S	2	9	16	23	30

September
M		7	14	21	28
T	1	8	15	22	29
W	2	9	16	23	30
T	3	10	17	24	
F	4	11	18	25	
S	5	12	19	26	
S	6	13	20	27	

October
M		5	12	19	26
T		6	13	20	27
W		7	14	21	28
T	1	8	15	22	29
F	2	9	16	23	30
S	3	10	17	24	31
S	4	11	18	25	

November
M	2	9	16	23	30
T	3	10	17	24	
W	4	11	18	25	
T	5	12	19	26	
F	6	13	20	27	
S	7	14	21	28	
S	1	8	15	22	29

December
M		7	14	21	28
T	1	8	15	22	29
W	2	9	16	23	30
T	3	10	17	24	31
F	4	11	18	25	
S	5	12	19	26	
S	6	13	20	27	

12

January
M		5	12	19	26
T		6	13	20	27
W		7	14	21	28
T	1	8	15	22	29
F	2	9	16	23	30
S	3	10	17	24	31
S	4	11	18	25	

February
M	2	9	16	23	
T	3	10	17	24	
W	4	11	18	25	
T	5	12	19	26	
F	6	13	20	27	
S	7	14	21	28	
S	1	8	15	22	29

March
M	1	8	15	22	29
T	2	9	16	23	30
W	3	10	17	24	31
T	4	11	18	25	
F	5	12	19	26	
S	6	13	20	27	
S	7	14	21	28	

April
M		5	12	19	26
T		6	13	20	27
W		7	14	21	28
T	1	8	15	22	29
F	2	9	16	23	30
S	3	10	17	24	
S	4	11	18	25	

May
M	3	10	17	24	31
T	4	11	18	25	
W	5	12	19	26	
T	6	13	20	27	
F	7	14	21	28	
S	1	8	15	22	29
S	2	9	16	23	30

June
M		7	14	21	28
T	1	8	15	22	29
W	2	9	16	23	30
T	3	10	17	24	
F	4	11	18	25	
S	5	12	19	26	
S	6	13	20	27	

July
M		5	12	19	26
T		6	13	20	27
W		7	14	21	28
T	1	8	15	22	29
F	2	9	16	23	30
S	3	10	17	24	31
S	4	11	18	25	

August
M	2	9	16	23	30
T	3	10	17	24	31
W	4	11	18	25	
T	5	12	19	26	
F	6	13	20	27	
S	7	14	21	28	
S	1	8	15	22	29

September
M		6	13	20	27
T		7	14	21	28
W	1	8	15	22	29
T	2	9	16	23	30
F	3	10	17	24	
S	4	11	18	25	
S	5	12	19	26	

October
M		4	11	18	25
T		5	12	19	26
W		6	13	20	27
T		7	14	21	28
F	1	8	15	22	29
S	2	9	16	23	30
S	3	10	17	24	31

November
M	1	8	15	22	29
T	2	9	16	23	30
W	3	10	17	24	
T	4	11	18	25	
F	5	12	19	26	
S	6	13	20	27	
S	7	14	21	28	

December
M		6	13	20	27
T		7	14	21	28
W	1	8	15	22	29
T	2	9	16	23	30
F	3	10	17	24	31
S	4	11	18	25	
S	5	12	19	26	

13

January
```
M    4 11 18 25
T    5 12 19 26
W    6 13 20 27
T    7 14 21 28
F  1 8 15 22 29
S  2 9 16 23 30
S  3 10 17 24 31
```

February
```
M  1  8 15 22 29
T  2  9 16 23
W  3 10 17 24
T  4 11 18 25
F  5 12 19 26
S  6 13 20 27
S  7 14 21 28
```

March
```
M     7 14 21 28
T  1  8 15 22 29
W  2  9 16 23 30
T  3 10 17 24 31
F  4 11 18 25
S  5 12 19 26
S  6 13 20 27
```

April
```
M    4 11 18 25
T    5 12 19 26
W    6 13 20 27
T    7 14 21 28
F  1 8 15 22 29
S  2 9 16 23 30
S  3 10 17 24
```

May
```
M  2  9 16 23 30
T  3 10 17 24 31
W  4 11 18 25
T  5 12 19 26
F  6 13 20 27
S  7 14 21 28
S  1 8 15 22 29
```

June
```
M     6 13 20 27
T     7 14 21 28
W  1  8 15 22 29
T  2  9 16 23 30
F  3 10 17 24
S  4 11 18 25
S  5 12 19 26
```

July
```
M    4 11 18 25
T    5 12 19 26
W    6 13 20 27
T    7 14 21 28
F  1 8 15 22 29
S  2 9 16 23 30
S  3 10 17 24 31
```

August
```
M  1  8 15 22 29
T  2  9 16 23 30
W  3 10 17 24 31
T  4 11 18 25
F  5 12 19 26
S  6 13 20 27
S  7 14 21 28
```

September
```
M     5 12 19 26
T     6 13 20 27
W     7 14 21 28
T  1  8 15 22 29
F  2  9 16 23 30
S  3 10 17 24
S  4 11 18 25
```

October
```
M  3 10 17 24 31
T  4 11 18 25
W  5 12 19 26
T  6 13 20 27
F  7 14 21 28
S  1 8 15 22 29
S  2 9 16 23 30
```

November
```
M     7 14 21 28
T  1  8 15 22 29
W  2  9 16 23 30
T  3 10 17 24
F  4 11 18 25
S  5 12 19 26
S  6 13 20 27
```

December
```
M     5 12 19 26
T     6 13 20 27
W     7 14 21 28
T  1  8 15 22 29
F  2  9 16 23 30
S  3 10 17 24 31
S  4 11 18 25
```

14

January
```
M    3 10 17 24 31
T    4 11 18 25
W    5 12 19 26
T    6 13 20 27
F    7 14 21 28
S  1 8 15 22 29
S  2 9 16 23 30
```

February
```
M     7 14 21 28
T  1  8 15 22 29
W  2  9 16 23
T  3 10 17 24
F  4 11 18 25
S  5 12 19 26
S  6 13 20 27
```

March
```
M     6 13 20 27
T     7 14 21 28
W  1  8 15 22 29
T  2  9 16 23 30
F  3 10 17 24 31
S  4 11 18 25
S  5 12 19 26
```

April
```
M    3 10 17 24
T    4 11 18 25
W    5 12 19 26
T    6 13 20 27
F    7 14 21 28
S  1 8 15 22 29
S  2 9 16 23 30
```

May
```
M  1  8 15 22 29
T  2  9 16 23 30
W  3 10 17 24 31
T  4 11 18 25
F  5 12 19 26
S  6 13 20 27
S  7 14 21 28
```

June
```
M     5 12 19 26
T     6 13 20 27
W     7 14 21 28
T  1  8 15 22 29
F  2  9 16 23 30
S  3 10 17 24
S  4 11 18 25
```

July
```
M    3 10 17 24 31
T    4 11 18 25
W    5 12 19 26
T    6 13 20 27
F    7 14 21 28
S  1 8 15 22 29
S  2 9 16 23 30
```

August
```
M     7 14 21 28
T  1  8 15 22 29
W  2  9 16 23 30
T  3 10 17 24 31
F  4 11 18 25
S  5 12 19 26
S  6 13 20 27
```

September
```
M    4 11 18 25
T    5 12 19 26
W    6 13 20 27
T    7 14 21 28
F  1 8 15 22 29
S  2 9 16 23 30
S  3 10 17 24
```

October
```
M  2  9 16 23 30
T  3 10 17 24 31
W  4 11 18 25
T  5 12 19 26
F  6 13 20 27
S  7 14 21 28
S  1 8 15 22 29
```

November
```
M     6 13 20 27
T     7 14 21 28
W  1  8 15 22 29
T  2  9 16 23 30
F  3 10 17 24
S  4 11 18 25
S  5 12 19 26
```

December
```
M    4 11 18 25
T    5 12 19 26
W    6 13 20 27
T    7 14 21 28
F  1 8 15 22 29
S  2 9 16 23 30
S  3 10 17 24 31
```

ZEROS TO ZILLIONS

AND MILLIONS TO BILLIONS

Large numbers can be very confusing:

An American million is the same as the British million and has six zeros

An American billion is the same as a British thousand million and has nine zeros

An American trillion is the same as a British billion and has 12 zeros

An American quadrillion is the same as a British thousand billion and has 15 zeros

An American quintillion is the same as a British trillion and has 18 zeros

An American sextillion is the same as a British thousand trillion as has 21 zeros

An American septillion is the same as a British quadrillion and has 24 zeros

An American octilion is the same as a British thousand quadrillion and has 27 zeros

An American nonillion is the same as a British quintillion and has 30 zeros

An American decillion is the same as a British thousand quintillion and has 33 zeros

The highest generally accepted named number is the centillion. In the United States a centillion is one followed by 303 zeros. In Britain it is one followed by 600 zeros.

The number one is followed by 100 zeros – 10 duotrigintillion in America or 10,000 sexdecillion in Britian – is universally known as a googol, a term devised by the late Dr Edward Kasner, an American maths guru who also gave us the googolplex, which is one followed by a googol of zeros.

In dealing with large numbers, serious mathematicians use the notation of ten raised to various powers to eliminate a profusion of zeros so that 100 is 10^2, 1000 is 10^3, 1,000,000 is 10^6, 1,000,000,000 is 10^9 and so on.

THE BIG TIME

Who wants to be a millionaire? I don't – if only because it would take me so long to count my fortune. Had I £1,000,000 in £1 notes and were I to count them out at the rate of 60 notes a minute, even working an eight-hour day, five days a week, it would still take me nearly seven weeks to finish the job.

No, it's hard work counting one's millions. Keeping an eye on them isn't easy either. I once met an American billionaire who told me that if his billion was piled up in $1 notes, the tower of money would be some 125 miles high. He had ambitions to become an American trillionaire, when his pile of banknotes would reach halfway to the moon.

I like big numbers, but not *that* big. Now we've nearly reached the end of the book I'd better introduce you to my favourites. The first is 100001. I like it because it looks good – and has style.

Take 100001 and multiply it by *any* five-digit number and you will get a ten-digit answer made up of the five digits of the multiplier repeated twice:

```
    100001
  ×  85246
8524685246

    100001
  ×  96314
9631496314

    100001
  ×  32158
3215832158

    100001
  ×  44444
4444444444

    100001
  ×  12121
1212112121
```

15873 isn't as superficially attractive as 100001, but multiply it by seven and multiples of seven and watch what happens:

```
15873 ×   7 = 111111
15873 ×  14 = 222222
15873 ×  21 = 333333
15873 ×  28 = 444444
15873 ×  35 = 555555
15873 ×  42 = 666666
15873 ×  49 = 777777
15873 ×  56 = 888888
15873 ×  63 = 999999
15873 ×  70 = 1111110
15873 ×  77 = 1222221
15873 ×  84 = 1333332
15873 ×  91 = 1444443
15873 ×  98 = 1555554
15873 × 105 = 1666665
15873 × 112 = 1777776
```

```
15873 × 119 = 1888887
15873 × 126 = 1999998
15873 × 133 = 2111109
15873 × 140 = 2222220
15873 × 147 = 2333331
15873 × 154 = 2444442
15873 × 161 = 2555553
15873 × 168 = 2666664
15873 × 175 = 2777775
```

37037 is another favourite. It has infinite possibilities. I'll multiply it by every number from one to 27 and you go on from there. I promise you you're in for a treat.

```
37037 ×  1 = 37073
37037 ×  2 = 74074
37037 ×  3 = 111111
37037 ×  4 = 148148
37037 ×  5 = 185185
37037 ×  6 = 222222
37037 ×  7 = 259259
37037 ×  8 = 296296
37037 ×  9 = 333333
37037 × 10 = 370370
37037 × 11 = 407407
37037 × 12 = 444444
37037 × 13 = 481481
37037 × 14 = 518518
37037 × 15 = 555555
37037 × 16 = 592592
37037 × 17 = 629629
37037 × 18 = 666666
37037 × 19 = 703703
37037 × 20 = 740740
37037 × 21 = 777777
37037 × 22 = 814814
37037 × 23 = 851851
37937 × 24 = 888888
```

```
37037 × 25 = 925925
37037 × 26 = 962962
37037 × 27 = 999999
```

The joy of numbers is that they are full of surprises. The most remarkable are often the least remarkable-looking and you can come upon them quite by chance.

142857 is my fourth great favourite. Lewis Carroll first ran across it when he divided it by seven:

$$1 \div 7 = 0.142857142857142857142857$$

Taking the minimum stretch of digits before they are repeated, Carroll then discovered:

```
142857 × 1 = 142857
142857 × 2 = 285714
142857 × 3 = 428571
142857 × 4 = 571428
142857 × 5 = 714285
142857 × 6 = 857142
```

In every answer the same digits appear, though starting with a different one each time.

Now looks what happens when you multiply 142857 by seven:

$$142857 \times 7 = 999999$$

When you multiply by eight or more the pattern reappears, though you have to make an additional calculation to get the proper sequence:

$$142857 \times 8 = 1142856$$

Take the first digit and add it to the rest and you get back to 142857 again

$$142856 + 1 = 142857$$

As the multipliers get larger, the pattern persists, though after a while you have to add the first two digits to the rest to keep the sequence going:

$142857 \times 16 = 2285712$ and
$\qquad 285712 + 2 = 285714$
$142857 \times 29 = 4142853$ and
$\qquad 142853 + 4 = 142857$
$142857 \times 34 = 4857138$ and
$\qquad 857138 + 4 = 857142$
$142857 \times 51 = 7285707$ and
$\qquad 285707 + 7 = 285714$
$142857 \times 64 = 9142848$ and
$\qquad 142848 + 9 = 142857$
$142857 \times 89 = 12714273$ and
$\qquad 714273 + 12 = 714285$
$142857 \times 113 = 16142841$ and
$\qquad 142841 + 16 = 142857$
$142857 \times 258 = 36857106$ and
$\qquad 857106 + 36 = 857142$
$142857 \times 456 = 65142792$ and
$\qquad 142792 + 65 = 142857$
$142857 \times 695 = 99285615$ and
$\qquad 285615 + 99 = 285714$

Come what may, seven and its multiples are a law unto themselves:

$142857 \times 28 = 3999996$ and
$\qquad 999996 + 3 = 999999$
$142857 \times 49 = 6999993$ and
$\qquad 999993 + 6 = 999999$
$142857 \times 63 = 8999991$ and
$\qquad 999991 + 8 = 999999$
$142857 \times 84 = 11999988$ and
$\qquad 999988 + 11 = 999999$
$142857 \times 203 = 28999971$ and
$\qquad 999971 + 28 = 999999$

What's more, watch what happens when you try multiplying the digits by seven on its own and then

179

dividing the result by nine:

$$142587 \times 7 = 999999$$
and $999999 \div 9 = 111111$

$$285714 \times 7 = 1999998$$
and $1999998 \div 9 = 222222$

$$428571 \times 7 = 2999997$$
and $2999997 \div 9 = 333333$

$$571428 \times 7 = 3999996$$
and $3999996 \div 9 = 444444$

$$714285 \times 7 = 4999995$$
and $4999995 \div 9 = 555555$

$$857142 \times 7 = 5999994$$
and $5999994 \div 9 = 666666$

PRIME AND PERFECT

Numbers like 142857 and 37037 and 15873 and 100001 are my prime numbers. To me they're perfect – though technically, of course, they're neither prime nor perfect and I must end on an accurate note.

A 'prime' number is a whole number which cannot be divided by any other whole number, except one, without leaving a remainder. The primes under 100 are: 2, 3, 5, 7, 11, 13, 17, 19, 23, 29, 31, 37, 41, 43, 47, 53, 59, 61, 67, 71, 73, 79, 83, 89 and 97. The highest known prime number is $2^{19937}-1$ (a number of 6,002 digits of which the first five are 43,154 and the last three, 471) received by the American Mathematical Society in March 1971, and calculated on an IBM 360/91 computer in 39 minutes 26.4 seconds.

A 'perfect', number is one that equals the sum of all its divisors, except itself – like six $(1 + 2 + 3 = 6)$ or 28 $(1 + 2 + 4 + 7 + 14 = 28)$. These are the top 20:

	FORMULA	NUMBER	NUMBER OF DIGITS
1	$2^1 (2^2 - 1)$	6	1
2	$2^2 (2^3 - 1)$	28	2
3	$2^4 (2^5 - 1)$	496	3
4	$2^6 (2^7 - 1)$	8,128	4
5	$2^{12} (2^{13} - 1)$	33,550,336	8
6	$2^{16} (2^{17} - 1)$	8,589,869,056	10
7	$2^{18} (2^{19} - 1)$	137,438,691,328	12
8	$2^{30} (2^{31} - 1)$	2,305,843,008,139,952,128	19
9	$2^{60} (2^{61} - 1)$		37
10	$2^{88} (2^{89} - 1)$		54
11	$2^{106} (2^{107} - 1)$		65
12	$2^{126} (2^{127} - 1)$		77
13	$2^{520} (2^{521} - 1)$		314
14	$2^{606} (2^{607} - 1)$		366
15	$2^{1,278} (2^{1,279} - 1)$		770
16	$2^{2,202} (2^{2,203} - 1)$		1,327
17	$2^{2,280} (2^{2,281} - 1)$		1,373
18	$2^{3,216} (2^{3,217} - 1)$		1,937
19	$2^{4,252} (2^{4,253} - 1)$		2,561
20	$2^{4,422} (2^{4,423} - 1)$		2,663

I feel I should end with the highest known perfect number. It is $(2^{19937} - 1) \times 2^{19936}$. It begins with 931, ends with 656 and runs to 12,003 digits in all. To give it to you in full I'm afraid I'd need another book, so I hope you'll understand if I keep it up my sleeve till then.

ANSWERS

Acidimeters to Zymometers

1 Gradients. **2** Calculation. **3.** Small electric currents. **4.** Water intake. **5.** Salinity. **6.** Specific gravity of urine. **7** Strength of silver solutions. **8** Polarization of light. **9** Vibrations. **10** Lung capacity.

Measure Miscellany

1 Measures eight magnums of champagne. **2** Measures electrical capacitance. **3** Measures 24 sheets of paper. **4** Measures electric current. **5** Measures a hundredth part of a chain. **6** Measures 202 yards. **7** Measures plane angle. **8** Measures frequency. **9** Measures inductance. **10** Measures 6·4 pints of champagne/brandy. **11** Measures force. **12** Measures electric charge. **13** Measures work, energy. **14** Measures power. **15** Measures thermodynamic pressure. **16** Measures electrical resistance. **17** Measures 480 sheets of paper. **18** Measures four inches. **19.** Measures 6076.1003 feet. **20.** Measures 120 fathoms.

Quizculation

1 1066×13	$= 13858$
2 13858×366	$= 5072028$
3 $5072028 - 28$	$= 5072000$

4 5072000 ÷ 1000	= 5072
5 5072 − 6	= 5066
6 5066 − 1066	= 4000
7 4000 ÷ 100	= 40
8 40 − 39	= 1

Blotting your copy book

```
   289
+  764
  1053
```

Make Me a Number

a) 6729 and 13458
6729 × 2 = 13458

b) 148 + 35 = 1
296   70

One in Three

1 Seventieth. **2** Seventh. **3** Fourteenth. **4** Thirtieth. **5** Fortieth. **6** Ninth. **7** Fifty-fifth. **8** Forty-fifth. **9** First. **10** Fourth. **11** Eighth. **12** Tenth. **13** Eleventh. **14** Twenty-fifth. **15** Thirty-fifth. **16** Thirteenth. **17** Second. **18** Fifteenth. **19** Sixth. **20** Third. **21** Twelfth. **22** Fifth. **23** Twentieth. **24** Fiftieth. **25** Sixtieth.

Movie Numbers

1 *Fahrenheit 451.* **2** *Eight and a half.* **3** Dirk Bogarde. **4** Sir William Walton. **5** *The Seven Samurai* and *The Magnificent Seven.* **6** *Five Graves to Cairo.* **7** *Forty-ninth Parallel.* **8** Yul Brynner. **9** *The Inn of the Sixth Happiness.* **10** Michael York, Frank Finlay, Oliver Reed, Richard Chamberlain. **11** Graham Greene. **12** *One Million Years BC.* **13** Arthur C

Three in Three

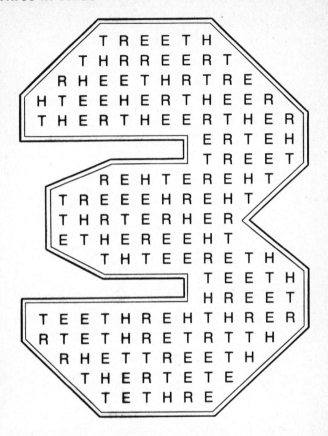

```
    T R E E T H
   T H R R E E R T
  R H E E T H R T R E
 H T E E H E R T H E E R
T H E R T H E E R T H E R
           E R T E H
           T R E E T
      R E H T E R E H T
   T R E E E H R E H T
   T H R T E R H E R
   E T H E R E E H T
      T H T E E R E T H
              T E E T H
              H R E E T
   T E E T H R E H T H R E R
   R T E T H R E T R T T H
   R H E T T R E E T H
      T H E R T E T E
      T E T H R E
```

Clarke: *2001: A Space Odyssey.* **14** *The Four Feathers.* **15** Kenneth More. **16** *The Wild One.* **17** *55 Days at Peking.* **18** *Twelve O'Clock High.* **19** Jules Verne: *Around the World in Eighty Days.* **20** *The Seventh Seal* by Ingmar Bergman.

Eleven Up

The lines should be drawn as shown on page 186.

Eight of the Best

1 Francis Bacon. **2** Euclid. **3** Pascal. **4** Bertrand Russell. **5** Carl Sanburg. **6** Sydney Smith. **7** Mae West. **8** Alfred North Whitehead.

Eleven Up

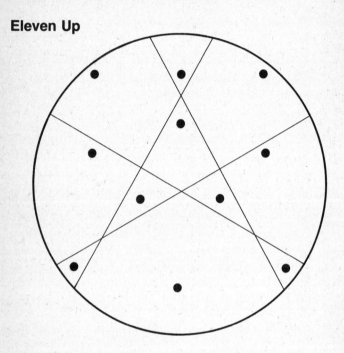

Three tricky squares

The common total for each of the rows, columns and diagonals in each of the three squares is five. You obtain this by adding two of the digits in each line and subtracting the third.

One eon?

one	eon
two	tow

three	ether
four	flour
five	verify
six	xis
seven	evens
eight	weight
nine	inner
ten	net
eleven	leavened
twelve	wavelet
thirteen	tethering
fourteen	counterfeit
fifteen	stiffener
sixteen	existent
seventeen	retentiveness
eighteen	heightened
nineteen	internecine
twenty	noteworthy
thirty	thirsty
thirty-one	retinopathy
thirty-six	thyrotoxicosis
thirty-seven	hypersensitivity
thirty-nine	interchangeability
forty	frosty
forty-one	confectionery
forty-nine	confectionery
fifty	stiffly
fifty-one	affectionately
fifty-nine	inefficiently
sixty	sexuality
sixty-three	heterosexuality
seventy	sensitively
eighty	weighty
eighty-one	homogeneity
eighty-nine	interchangeability
ninety	intently
ninety-one	conveniently

ninety-eight   interchangeability
ninety-nine   inconveniently
one hundred   undernourished

Note that 'confectionery' provides both forty-one and forty-nine; and that 'interchangeability' provides thirty-nine, eighty-nine, and ninety-eight.

The Joy of Sex

1 Sexennarian. **2** Sextain. **3** Sextuple. **4** Sexcuple. **5** Sextoncy. **6** Sexto. **7** Sexagene. **8** Sexdigital. **9** Sexfoil. **10** Sexadecimal. **11** Sexvirate. **12** Sextipartition. **13** Sexennial. **14** Sexfid. **15** Sextans. **16** Sexavalent. **17** Sexagenary. **18** Sextur. **19** Sexdigit. **20.** Sexmillenary. **21** Sextile. **22** Sexennium. **23** Sexious. **24** Sexen. **25** Sexagecuple.

Letters Count

a)
```
  4 6 3 8 8
  1 4 2 8 8
  2 3 8 8 8
  ─────────
  8 4 5 6 4
```

b)
```
  2 2 9 1 2 4 4
  2 2 2 8 3 4 4
  2 2 3 1 2 4 4
  ─────────────
  6 7 5 0 8 3 2
```

c)
```
  2 7 2
  4 7 2
  ─────
  7 4 4
```

d)
```
  5 5 2 5
  5 5 6 6
  5 2 2 5
  5 6 9 6
  ───────
  2 2 0 1 2
```

Count Up

a)
```
  173              85
+   4    and    + 92
─────           ─────
  177             177
```

b) Hold the sum in front of a looking-glass and you'll see it makes sense.

188

c)
```
    1
    1
    1
 +11
 ──
  14
```

d)
```
   15
   36
   47
 +  2
 ───
  100
```

e) $9 + \dfrac{99}{99} = 10$ $\qquad\qquad$ $\dfrac{99}{9} - \dfrac{9}{9} = 10$

f) $99\,\dfrac{9}{9} = 100$

Palindromes

Noon

Missing Numbers

a 5. **b** 5. **c** 63. **d** 27. **e** 3. **f** 49.

Challenging Numbers

1 £100. **2** 24. **3** 15. **4** Two. **5** Two miles uphill, four miles downhill, and three miles on the level. **6** Plus three and plus five or minus three and minus five. **7** 22/3. **8** 35. **9** 34 feet. **10** 54 pounds. **11** 20 per cent. **12** Bill. **13** 4. **14** 48.

A Question of Age

a) Dick is one-year old.

Chinese Challenge

$$\overline{\overline{\equiv}}\ \overline{\overline{\equiv}} = 4$$

What have we here?

This is Lewis Carroll's answer:
The Whole, – is Man.
The Parts are as follows:
A large Box – The Chest.
Two lids – The Eye Lids.
Two Caps – The Knee Caps.
Three established Measures – The nails, hands and feet.
A great number of articles a Carpenter cannot do without – nails.
A couple of good Fish – The Soles of the Feet.
A number of a small tribe – the Muscles. (Mussels)
Two lofty Trees – the Palms (of the hands).
Fine Flowers – Two lips (Tulips), and Irises.
The fruit of an indigenous Plant – Hips.
A handsome Stag – The Heart (Hart)
Two playful Animals – The Calves
A number of a smaller and less tame herd – The Hairs (Hares)
Two Halls, or Places of Worship – The Temples.
Some Weapons of Warfare – The arms, and Shoulder blades.
Many Weathercocks – The veins. (Vanes)
The Steps of an Hotel – The Insteps (Inn-Steps)
The House of Commons on the eve of a Dissolution – Eyes, and Nose (Ayes and Noes)
Two Students or Scholars – The pupils of the Eye.
Some Spanish Grandees – The Tendons. (Ten Dons)

Fractions Cipher
I KNEW YOU COULD WELL DONE

Twishum Cipher
MI5 HERE I COME

Map Reference Ciphers

a) 01474319173907136313492739072733136367
6919

b) 5903076333340969531363471349071913 6359

c) 63470447390963074969633370734